Maria Bruno

Pets, Professors, and Politicians

Pets, Professors, and Politicians

The Founding and Early Years
of the Atlantic Veterinary College

Island Studies Press
Atlantic Veterinary College
Charlottetown

2004

Pets, Professors, and Politicians:
The Founding and Early Years of the Atlantic Veterinary College
© 2004 by Marian Bruce
ISBN 0-919013-43-0

Cover image: *Night Group*, by Lindee Climo
Spot illustrations: Jay Ryan
Editing: Edward MacDonald
Photography: UPEI Photo Services
Design: UPEI Graphics
Printing: Friesens Book Division

Library and Archives Canada Cataloguing in Publication

Bruce, Marian
    Pets, professors, and politicians : the founding and early years of the Atlantic Veterinary College / Marian Bruce.

Co-published by the Atlantic Veterinary College.
Includes bibliographical references and index.
ISBN 0-919013-43-0

    1. University of Prince Edward Island. Atlantic Veterinary College—History. 2. Veterinary medicine—Study and teaching (Higher)—Prince Edward Island—Charlottetown—History. I. University of Prince Edward Island. Atlantic Veterinary College II. Title.

SF756.37.C32A85 2004   636.089'71'1   C2004-903051-5

Island Studies Press
University of Prince Edward Island
Charlottetown, Prince Edward Island
Canada C1A 4P3
www.upei.ca

Atlantic Veterinary College
University of Prince Edward Island
Charlottetown, Prince Edward Island
Canada C1A 4P3
www.upei.ca

Printed in Canada

To
Dr. R. G. "Reg" Thomson,
inaugural and indefatigable
dean of the
Atlantic Veterinary College,
and to all those for whom he built this school,
its graduates, past and future.

Contents

Foreword

L et me begin by stating what this book is not.
This is not a typical institutional history. It not a weighty academic tome that peers over its glasses at the class (though it keeps its audience in mind). It does not wade ponderously through a tide of comparative statistical and sociological analysis (though you will find some of both here). It is not a relentlessly comprehensive history, amassing factoid after factoid for the sake of the record (though it is alive with telling details). It is not a self-serving sermon of praise (though it finds much to compliment and much to celebrate). It is none of these things because that is not its intent.

Pets, Professors, and Politicians: The Founding and Early Years of the Atlantic Veterinary College is a lively, intelligent, probing re-telling of the history of the Atlantic Veterinary College. The ink on that history is still wet, and yet it is not too soon to try to capture the taste and texture of its past, especially now, when memory can still inform, enrich, and interpret it. While it may be too soon to trace some of the larger contours of the story, in other respects, it is almost too late.

And so we have Marian Bruce's book. It is a fascinating account, and not just to those whose lives and careers have been bound up with the AVC. I can think of at least three reasons for this. First, the book brims with interesting people doing interesting things: veterinarians, politicians, administrators, teachers, students, technicians, researchers—even a few animals. Second, the tale of its founding, a saga stretching over fifteen years, with more twists and turns than a suspense novel, is a must-read for anyone interested in the politics of higher education in Canada. Third, the history of the AVC's first two decades is a story surpassingly well told. The teller, Marian Bruce, combines the narrative gifts of a born storyteller with the skilled journalist's sense that the "story" resides in people as much as events. Her abiding concern for character and motivation parallels the brisk chronicle of bricks and mortar, programs and products, trends and impacts. Institutional history is often as instructive as this and occasionally as judicious, but it is seldom so colourful.

After twenty years, the story of the Atlantic Veterinary College continues to unfold. Thanks to this book, the telling has begun.

Edward MacDonald
Department of History, University of Prince Edward Island

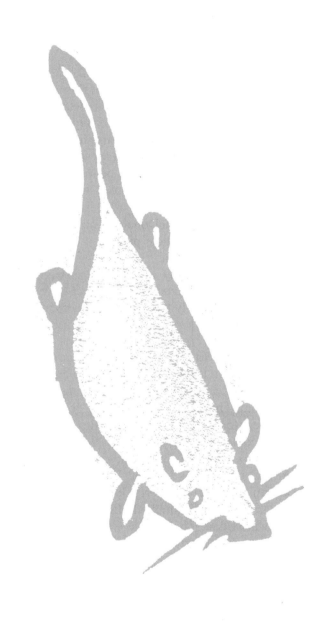

Acknowledgements

Many people contributed to this book. First and foremost, though, I am indebted to members of the AVC history committee, who guided me with a loose rein and helped make this project a joy from beginning to end: Dean Tim Ogilvie, former dean Larry Heider, assistant dean Mel Gallant, Dr. Bob Curtis, Dr. Bud Ings, Natalie Fraser, and Erin Gray. They gave me the freedom to record the story of AVC the way I saw it, as a relatively objective observer. A special thanks to Dean Ogilvie, who wanted a history that people might actually read, and who never shrank from any contentious issues it explored.

The committee was fortunate enough to engage, as editor, UPEI history professor Dr. Edward MacDonald, who also happens to be a superb writer. Throughout the whole process, I relied, with complete confidence, on his good judgement.

Thanks are due to the many people interviewed for the book. Among them was Helen Thomson of Woodstock, Ontario, who opened her home to me and generously shared her memories of her late husband, Dr. Reg Thomson, and of the family's sojourn in Prince Edward Island.

Thanks also to Simon Lloyd and Leo Cheverie of Special Collections at UPEI's Robertson Library, and Marilyn Bell of the provincial Public Archives and Records Office, for helping unearth valuable documented material. Thanks, too, to Shelley Ebbett of UPEI's Photo Services Department for all her work finding these photos.

A number of people read the manuscript, or portions of it, before publication: members of the history committee, University of Prince Edward Island President Wade MacLauchlan, Dr. Peter Meincke, Dr. Ray Long, Dr. Jim Bellamy, Dr. Gerry Johnson, Dr. Wendell Grasse, Dr. Andrew Peacock, Alex Campbell, Bob Nutbrown, Larry Durling, and Barry MacMillan. Their advice, when I followed it, helped improve the final product greatly. Mel Gallant, assistant dean at AVC, read several drafts of the manuscript. His eagle eye and astonishing command of the fine points of the language saved me from many a gaffe. After all that, any errors, omissions, or other inadequacies in the book are my responsibility.

Marian Bruce

1: Hard Times

One fall morning in 1948, Albert "Bud" Ings boarded a train in Charlottetown, paid the conductor a quarter for a pillow, and started on a two-day journey to Guelph, Ontario, sitting up all night because he couldn't afford to pay $5 for a sleeping berth. Bud, a farm boy from Mt. Herbert, Prince Edward Island, was on a mission: he planned to study veterinary medicine at the Ontario Veterinary College (OVC) in Guelph. He was one of the first Islanders to do so.

The cost of a year at OVC seems slight in retrospect — $100 for tuition, $50 for books, and $9 a week for room and board — but a boy from the farm in those days didn't carry much spare change. There was no student aid, and the money for Bud's college education had to be raised from the meagre resources of the family farm, summer employment, and part-time work at the college (fifty cents an hour) during the school year. That did not deter Bud and two other local boys, Hammond Kelly and Russell Furness. Veterinary medicine had become, Bud recalled, "an obsession with some of us young fellows."

That obsession did not spring from dreams of a profession in which a young man could make an easy dollar. Veterinarians throughout the Atlantic region treated mostly farm animals, horses being the most valuable — and valued. That meant days and nights of farm calls, slogging through mud to the axles in the spring, and plowing through snowdrifts in winter. If the veterinarian was lucky, he might have an extra $5 in his pocket when he left for home.

What motivated Bud Ings was his experience of seeing, and hearing about, the loss of cattle and horses for want of professional help. His family had a mixed farm with a large herd of purebred Ayrshire cattle and three or four Clydesdale horses. "When I was in my early teens, there was only one veterinarian we could call on," he recalled. "The man was so busy he couldn't cover all the territory required. He was rushed off his feet, so he was difficult to get hold of. It was kind of sad."

Albert E. Ings, Sr. (age 81), with his grandson, Albert E. Ings, Jr. (age 18), in 1944

The Island boys who left for OVC had also received much encouragement — although no financial aid — from Walter Jones, a neighbour and close friend of the Ings family. He wielded considerable influence with his neighbours, as he happened to be premier of Prince Edward Island, and, as owner of a herd of prize Holsteins and a proponent of a scientific approach to agriculture, he spoke from first-hand experience about the difficulties of obtaining qualified veterinary help.

In those days, most farmers on the Island, as in other parts of the Atlantic provinces, did their own doctoring, or called in a neighbour reputed to have a knack for treating animals. The lay veterinarians, usually called "handymen" or the less complimentary "quacks," generally had some experience in castrating livestock and treating sick animals, but had little or no formal education. Some used remedies passed down through generations; some had framed certificates hanging on the kitchen wall, attesting to their having completed correspondence courses from private schools in Ontario or the United States. The handymen often proved helpful in difficult calving cases or simple castration procedures. But farmers were suffering heavy losses from serious diseases — Bang's disease and mastitis in dairy cattle, anaemia in pigs, parasitic diseases in sheep and foxes, fowl typhoid in poultry. Sometimes the handymen helped; sometimes the treatment was worse than the disease.

Early Veterinary Education

Canada's oldest veterinary college began its life as a private school in Toronto in the 1860s. In response to concerns by Upper Canadian farmers' groups about the lack of qualified veterinary care, the Upper Canada Board of Agriculture persuaded Andrew Smith, a graduate of the Edinburgh Veterinary College, to set up a practice in Toronto, with a view to establishing a veterinary school. In 1861, advertisements in the local newspapers announced the opening of Smith's "Horse Infirmary and Veterinary Establishment," where "Veterinary Medicines of every description are constantly kept on hand — such as Physic, Diuretic, Cough Cordial, Tonic, Worm Balls ...."

In 1862, Smith began giving four-week courses in veterinary science to young farmers, and more advanced courses to young men wishing to obtain "diplomas in the veterinary art." In the next few years, he bought buildings and equipment and hired faculty for his school, known as the Upper Canada Veterinary School.

The first faculty member was Duncan McEachran, a fellow Scot and former classmate. McEachran and Smith disagreed over the issue of academic standards: McEachran thought Smith's were appallingly low. McEachran left the school after three years to begin teaching veterinary science in Montreal, and founded his own school there in 1875. It offered a three-year program in veterinary medicine, compared with a program of nine or ten months at the Toronto school. McEachran's school lasted for twenty-eight years, a respectable length of time considering the failure of most private veterinary schools in the United States in that era.

Meanwhile, the Toronto school continued to flourish. By the 1890s, close to four hundred students were enrolled at the institution, by then known as the Ontario Veterinary College. In 1908, when the Ontario government took over the college, the program was extended to three years. Ten years later, the course was lengthened to four years, and in 1949 to five years. The college was moved to Guelph in 1922.[1]

Farming in those days was becoming more specialized and more businesslike, and more and more farmers were starting to recognize the need for better-qualified veterinary help. The shortage existed across Canada. In 1945, the Prince Edward Island Veterinary Medical Association noted, "It has been estimated that we require 2,000 veterinarians in Canada to meet the demand of departmental and commercial services as well as to provide a sufficient number of veterinary practitioners. The number available is approximately 900. This is a serious condition of affairs, and it affects every human individual in the country. Should the number of veterinarians tend to decrease instead of increasing in the future, the result will be a tremendous economic loss through wastage by disease and death in livestock, and in the curtailment of public health services...to the detriment of the general public as a whole."[2]

In 1950, Prince Edward Island had six veterinarians, two in private practice and four in government service. Newfoundland had only one veterinarian, employed by the government. Farm service in Newfoundland was nonexistent before the mid-1960s. Distances between farms were so great, a veterinarian in private practice couldn't hope to make a living. "Farmers just muddled through on their own," said Dr. Alton Smith of Cupids, Newfoundland, one of the first Newfoundlanders to graduate from OVC. Nova Scotia had twenty-two veterinarians, thirteen of whom were in private practice. And New Brunswick had twenty-eight, sixteen of them in private practice.

To supply more veterinarians, and to accommodate returning veterans after the Second World War, class sizes at Canada's two veterinary colleges, OVC and the Université de Montréal's Faculté de médicine vétérinaire in Saint-Hyacinthe, Québec, were substantially increased. (Veterans who wanted to go to university received free tuition and a living allowance from the federal government, and they were flocking to the universities in droves.) Bud Ings' class at OVC had about 165 students, compared with the usual 50 or 60. As a result of that expansion, the output of veterinarians in Canada more than doubled between 1947 and 1950. As well, European-trained veterinarians helped fill the gap. But a chronic shortage continued to exist in the federal service and in some rural areas.[3]

Of course, the number of students accepted in veterinary col-

Early Veterinary Medicine in Canada

The horse was essential to survival in rural Prince Edward Island
(Public Archives of PEI)

leges in those days was quite different from the number who actually graduated. Ings recalled his first day in class, when a professor gave his standard welcome-to-OVC speech: "Look to the left of you," the professor intoned, "and look to the right of you. Because when it's time to graduate, one of you will not be here. We will weed you out like carrots!'" Dr. George Irving, who grew up in Rexton, New Brunswick, and enrolled at OVC in 1954, remembered the same warning — and the consequences of the weeding-out-like-carrots philosophy. He was one of three Atlantic Canadians in a class of sixty-five, about half of whom survived to graduation.

The monetary rewards at the end of that road were not great. Dr. Ross Ainslie of Halifax, who graduated from OVC the same year as Bud Ings, said the "quacks" undercut prices, and many farmers in that era didn't realize the value of paying for more professional care. Many simply couldn't afford to. "In the early 1950s, we were just coming out of a recession/depression. There wasn't a lot of money around. As a result, a farmer would have a look at a cow, call the vet, ask how much it was going to cost. Say it was $500 or $800 or whatever, they'd say, 'No, we're going to ship her.'"

To stabilize veterinarians' incomes and encourage farmers to use veterinary services, some provinces, including the Maritimes, established assistance programs. In the late 1930s and 1940s, Saskatchewan, Quebec, New Brunswick, and Nova Scotia set up systems assigning veterinarians to prescribed zones. In Nova Scotia, the province initially guaranteed a grant of $2,000 a year to the veterinarian, who then charged $2 or $3 a call regardless of distance travelled, plus the cost of medicines, surgery, and materials. Under New Brunswick's system, the veterinarian received a salary of $2,000 a year, plus eight cents a mile in travel expenses. The province also provided office space and paid for medicines, biologics, and surgical instruments. During daylight working hours, the veterinarian turned fees for service over to the government, but, after 5 P.M., he could do a little moonlighting with small-animal cases. (By 1959, when George Irving graduated, the salary had increased to $4,900; farmers paid a flat fee of $5 to the government, regardless of distance from the clinic or the time involved.) Newfoundland's Dr. Alton Smith set up its farm-service system in 1965, placing six or seven practitioners on salary to serve various parts of the province.

Dr. Alton N. Smith

Dr. George C. Fisher

Dr. J. T. "Joe" Akins

Dr. E. Errol I. Hancock

Four pioneers of early veterinary medicine in Atlantic Canada.
Dr. Alton Smith of Newfoundland and Labrador was the first Provincial Veterinarian (1965–84), pioneering veterinary programs and drafting animal health legislation in a province that was not yet oriented toward agriculture. Dr. George Fisher, originally from New Brunswick, served as PEI's Director of Veterinary Services, and was active regionally, nationally, and internationally as a practitioner, teacher, administrator, public educator, researcher, and policy developer. Dr. Joseph Akins joined the New Brunswick Department of Agriculture in 1937, establishing a farm service under the Livestock Branch, where he oversaw the development of a laboratory diagnostic service and drug supply depot, and in 1945 headed up a separate Veterinary Branch with a large animal clinic, which added surgical, radiological, and hospitalization facilities in 1959. Dr. Errol Hancock was responsible for setting up Nova Scotia's first animal pathology lab, in Truro, in 1937, and played a leading role in establishing one of the first veticare programs, as well as one of the first artificial breeding units, in Canada.

Premier Walter Jones set up a similar farm-service system in Prince Edward Island in 1950. By the time Bud Ings graduated from OVC, the Island government had hired four veterinarians from his class to work out of practices across the Island. In exchange for a subsidy of about $3,000 a year, they were required to keep their fees for farm calls down to $3, although the farmer paid for drugs and other costs.

Ings started his practice in Souris, then moved to Montague when the resident veterinarian there left for the United States. For a number of years, Ings was the only veterinarian in King's County, which meant making house calls thirty or forty miles in any one direction. "It was quite a chore," he recalled. "The set-up on the farm was that most calvings took place from March to June. There weren't too many emergencies in January and February, but in the spring of the year, when the roads got bad, that's when the problems came in. It would be nothing to work eighteen to twenty hours a day, because you'd just get home from Wood Islands at two o'clock in the morning, and a call would come in from East Point and you'd have to start out again, and maybe get home at five in the morning. It was rough."

In the early years, his practice consisted almost entirely of large animals. "If a dog got a broken leg, the odd person might spend $5 or $10 getting the leg repaired. But a cat — they wouldn't bother. I very seldom did any vaccinating or neutering on cats because there was a belief that you had to have kittens around the barn, and if a cat didn't have kittens, she wouldn't be any good to look for mice and rats."

When Ainslie graduated, he did a nine-month stint in a Tatamagouche, Nova Scotia, large-animal practice, and then moved to a clinic in Halifax. Even there, half his patients were horses and cattle, because his territory extended about a hundred miles east of the city and thirty miles west. "In this area, every little fishing village had three or four cows," he recalled. "When I came here first, they were still hauling milk wagons with horses. I saw the demise of that. Then there were light horses and race horses."

As the number of farms declined in the postwar years, and tractors, cars, and trucks gradually took over from horses, small animals began playing a larger and larger role in both rural and city practices. Improvements in drugs and equipment, as well as better forms of transportation and communication, made life easier for the veterinarian. However, veterinarians continued to be in short supply in some areas, particularly in the livestock industry.

In 1969, the Canadian Agricultural Services Co-ordinating Committee (CASCC) — composed of provincial and federal bureaucrats and the deans of colleges of agriculture and veterinary medicine — decided it needed some solid information on the state of the nation's veterinary manpower situation. Despite the opening of a new veterinary medical school that year at the University of Saskatchewan, the Western College of Veterinary Medicine, concern had been expressed about the shortage of veterinarians to serve the livestock industry in Canada and the lack of educational opportunities for would-be veterinarians.[4]

In 1970, Agriculture Canada commissioned a study to assess the need for veterinarians. The study, by Professor T. L. Jones, formerly Dean of OVC, and Dr. W. A. Moynihan, formerly of the federal Health of Animals branch, concluded that veterinary manpower in Canada was, in fact, insufficient. "New members joining the veterinary profession are being attracted to new fields of activity that are now beckoning the veterinarian," the authors stated. "As a consequence, farm practice and government service, both of which serve the livestock industry, tend to be placed in a lower priority of interest." These "new fields of activity" presumably referred to increases in small-animal cases, a situation that reflected Canada's rapid postwar transformation into an affluent, pet-owning, urban/suburban society.

The authors noted that between 1950 and 1970, the percentage of the profession engaged in food-animal service had declined to 69.2 per cent from 85.4 per cent. "The most feasible way to improve the prospects for veterinary service to meet the needs of the livestock industry is to increase the output of veterinary colleges to the point where all categories of employment would be filled," the report said.

The report recommended that the three universities expand their classes so as to graduate at least 100 more veterinarians a year — 65 from the University of Saskatchewan, compared with 34 in 1970; 60 from the Université de Montréal compared with 32 in 1970; and 100 from the University of Guelph compared with 53 in 1970.[5]

In fact, expansion plans had already begun. By 1974, OVC was accepting 120 students; the next year, Saint-Hyacinthe accepted 75, and the Western College contemplated increasing its graduates to 90 a year. However, as a result of the Jones-Moynihan report, the CASCC concluded in April 1971 that consideration should be given to establishing a fourth veterinary school.

Forecasting manpower needs in any field is not an exact science, and, at the time of the Jones-Moynihan report, there was a considerable divergence of opinion about veterinary manpower needs in Canada, and doubt in some quarters about the usefulness of such projections. What was less debatable was the student demand for education in veterinary medicine. In both Canada and the United States, it was easier to get into medical school than into a college of veterinary medicine. And Atlantic Canadians were especially handicapped, despite living in an area where land and sea animals dominated the economy. The entire Atlantic region was rarely allowed to place more than six students a year at OVC.

Dr. Don Rushton, who works out of a veterinary clinic in St. Margaret's Bay, Nova Scotia, was one of three Nova Scotia students accepted at Guelph in 1972. He had just graduated from Dalhousie University with a Bachelor of Science degree and had studied at the Nova Scotia Agricultural College in Truro, which, he said, gave him an advantage. "I had a lot of friends from various universities who tried to get into veterinary school and didn't, but went on to study human medicine," he said. "They're now MDs."

Some of the more determined Newfoundland students rejected by the Ontario and Saskatchewan colleges crossed the Atlantic to schools in England, Scotland, and Ireland. In a letter to OVC Dean Dennis Howell in 1975, Dr. Michael Bland, then president of the Newfoundland and Labrador Veterinary Medical Association, wrote, "It is disheartening to witness the extreme despondency among the potential veterinary students in the Newfoundland schools due to the fact that they have been almost totally ignored by the Canadian Veterinary Schools." Bland added that the Newfoundland students "have received no encouragement whatsoever from the Canadian Veterinary Medical Association, their inquiries frequently not even receiving acknowledgement." And it was not only students who were at a disadvantage without a regional college, Bland contended.

Dr. Ross Ainslie, with Abbey Barrett, in January 2004

Practising veterinarians in the region were unable to take continuing education courses because of the distance from the nearest college, and clients rarely could make use of the services a college could offer. Not surprisingly, Bland expressed the "wholehearted unanimous support" of his association for a fourth school in the east.

There was not, however, unanimity about the proposed college among veterinarians in other parts of Atlantic Canada. By the mid-1970s, Ross Ainslie, who was building up a fleet of clinics in Nova Scotia, had his doubts about the need for a fourth college. So did other Nova Scotian veterinarians, such as Ross Mitton and Gerald Sheehy. George Irving, who was employed as a salaried veterinarian in New Brunswick, eventually became a strong advocate for another college, but he initially thought that the best solution would be for OVC to boost its quota of students from the Atlantic region. Meanwhile, Bud Ings, who was phasing out his practice, had entered provincial politics. As Minister of Agriculture for Prince Edward Island, he was enthusiastically campaigning to have the proposed new school built in Charlottetown at Canada's newest university, the University of Prince Edward Island.

2: Looking for a Home

It was not an auspicious beginning for a new university. Some, of course, celebrated the birth of the University of Prince Edward Island. Others went into mourning. Some feared the newborn, or predicted its early demise. To ensure its survival, its guardians had to persuade Islanders to transcend family traditions and religious loyalties that had been etched into their bones for generations.

For a long time, the Island had been divided, mostly peacefully, along religious lines. There were Roman Catholic neighbourhoods and Protestant neighbourhoods, a Catholic hospital and a Protestant hospital, a Catholic orphanage and a Protestant orphanage, and a carefully observed tradition of alternating between the two faiths in many government appointments. Until the late 1960s, there were also two post-secondary institutions: the Roman Catholic St. Dunstan's University and the government-supported Prince of Wales College, technically nonsectarian but considered by many to be basically Protestant. Each had a long history and a fiercely loyal constituency.

In the late 1960s, the Island government led by Premier Alex Campbell decided it no longer made sense to support two competing universities in a province of 110,000 people. In April 1968, Premier Campbell bravely announced that the government planned to create one publicly funded institution, the University of Prince Edward Island. If the two existing institutions wanted to continue, they'd have to manage on their own.

Neither tried to. Much to the lingering dismay of some loyalists, the Prince of Wales property was converted to a technical college, and St. Dunstan's sold its buildings and grounds to the new university. Some faculty of both institutions declined to work for the upstart university, but many, including chemistry teacher Glenn Palmer, moved to the new campus. Palmer, who had taught at Prince of Wales for two years, was from Ontario and could therefore remain

Alex Campbell, Former Premier of PEI

somewhat dispassionate about such traumatic breaks with Island tradition.

Before UPEI opened its doors for the first time in 1969, Palmer recalled, there were dire predictions about its future. "There were suggestions that there was going to be blood on the floor," he said. "Protestants and Catholics could not possibly work in the same organization. I can recall a man at an open meeting saying that no Catholic family would send their children to UPEI. It was pretty amazing for someone coming in from Ontario."

There also was talk that, except for people who couldn't afford to go off-Island, Protestant students would continue to get their degrees at Acadia or Mount Allison universities, and the Catholics would continue to go to St. Francis Xavier. "We had just bought our house [in Charlottetown]," Palmer said, "and a colleague of mine at Prince of Wales said he was so sorry to hear that — obviously I hadn't understood how serious this all was and how disastrous it was going to be. I was going to be left with no job and no house."

Dr. Ron Baker, first president of the University of Prince Edward Island

This was the climate of uncertainty that Dr. Ronald Baker faced when he became the first president of the new university. In the beginning, he told an interviewer later, "I refused to have anything to do with the history of PWC or SDU. I just wouldn't listen. I very consciously kept telling people, 'It'll work out.'"[1]

Baker was no stranger to academic challenges. He went to UPEI from Simon Fraser University, where, as director of academic planning, he was mainly responsible for recruiting the first department heads in the mid-1960s. A year after Baker started work, Simon Fraser opened its doors on its mountain site with about 2,500 students and 350 faculty. At UPEI, besides dealing with — or perhaps determinedly ignoring — the cadavers of the two old institutions, Baker had to build a viable liberal arts institution, reassure the public and his own faculty that it would survive, and sort out the kinds of programs that would benefit both the Island and the new university.

He recalled being lobbied, sometimes at public meetings, to include all sorts of subjects that hadn't been taught previously on the Island. One of those subjects was agriculture. "People thought it would keep their children on the farm," he said.

It wasn't easy to keep young people on the farm in those days. Net farm income hadn't increased much since the end of the Second World War. By the time of the 1961 census, the number of Island farms had declined to 7,335 from 12,230. The farm population in that period fell by one-third, to 34,514 people.[2]

The farm crisis was not a new phenomenon on the Island, of course, nor was the idea of teaching young people about agriculture. At least as far back as 1900, Prince of Wales College had been pressured, off and on, to offer courses in "scientific agriculture." And, in 1922, the Carnegie Foundation had proposed that St. Dunstan's University trade in its liberal arts tradition for an agricultural school. St. Dunstan's administration rejected the idea of agriculture out of hand. So did President Baker.[3]

Baker knew that no new faculties of agriculture had been set up in Canada for decades, and existing ones were in trouble. "The idea that studying agriculture got students back to farms didn't work out. As John Kenneth Galbraith pointed out, he went to agriculture college to get off the farm, not to stay on it. Grads went into government, the aggie business, and so on. Children of successful, large-scale farmers tended to want business administration. Agriculture was increasingly becoming either marketing and business management or science in general, science that was happier in science faculties."

But Baker did hope to offer courses somehow related to the Island's farming tradition. It so happened that he had at least a nodding acquaintance with veterinary education, having kept tabs on negotiations for the Western College of Veterinary Medicine in the 1960s. For one thing, he knew that it was easier to get into medical school than into veterinary college.

Baker recalled that in the early 1970s, he began casually sounding out various people — some local veterinarians, farmers, other Maritime university presidents — about the possibility of a new veterinary college in the Atlantic region. He didn't encounter much enthusiasm. "The local vets were frightened of possible competition from a vet school. The farmers were worried that there would be 'too many women' who wouldn't be able to deliver a cow and would want to put their energy into pets." Neither Dalhousie University in

Dr. Herb MacRae, former principal of Nova Scotia Agricultural College

Halifax nor the University of New Brunswick in Fredericton showed much interest in hosting a veterinary school.

One Maritime institution Baker failed to consult was the Nova Scotia Agricultural College in Truro — presumably because he considered it little better than a vocational school "with courses in non post-secondary butchery, horse-shoeing, floristry and so on." At the time, the college, with an enrolment of fewer than four hundred students, offered only two years of a degree program, had no library building, no athletic facilities, and no university-level science faculty.

On the other hand, the college served as a preparatory school for some Atlantic Canadians who were accepted at veterinary school, had begun a major expansion program that was to include a library, athletic centre, laboratories, and an animal science building, and planned to acquire degree-granting status. The principal, Dr. Herb MacRae, had no difficulty seeing the logic of a marriage between his college and the proposed veterinary school. In fact, he considered it a duty. Agriculture and veterinary medicine had always worked closely together across Canada. Both programs were expensive to run because of the animal component, and the two schools could save money and enrich each other by sharing facilities. Dalhousie University, sixty miles away in Halifax, could provide support through its medical school, laboratories, and science courses.

Nevertheless, MacRae said, he was ambivalent about the prospect of a veterinary school on campus — and his staff were downright hostile to the idea. "This was never construed by the public to be honest on our part, and yet I can assure you that it was," he said, "and anybody on our staff would tell you this: there was great fear here about having the veterinary college, genuine fear on the part of the staff, as there was at UPEI — fear because of the huge amount of money it demanded, and how it would overshadow our regular programs.

"We were on the move then to becoming degree-granting, and to expanding into a four-year program, which would complement everything that would be done at the veterinary college, so it was all fitting together. But they were very much afraid of it. I can remember meeting with all our staff where we said, 'Look, there's going to be a fourth college of veterinary medicine, and we have an obligation to have it here.... Not that we really want it, but we have an obligation to have it.' For me to have said that publicly at the time, I would have been just laughed out of town. But that was the truth." Meanwhile, it was starting to look as though the new veterinary school would be moving in, whether MacRae and his staff wanted it or not.[4]

MacRae's recollection of events was that in October 1973, ministers and deputy ministers of Agriculture from the Atlantic provinces met in Truro to discuss the establishment of a fourth college in Truro. "I attended that meeting," Dr. MacRae said. "They all agreed that there should be a fourth college and it should be located here." The following August, Dr. Kenneth Wells, then Veterinary Director General for Agriculture Canada, visited Nova Scotia to discuss with officials of NSAC and the provincial government the establishment of a fourth college. From this visit, the Agricultural College officials inferred that federal Agriculture Minister Eugene Whelan favoured Truro as a site for the proposed college. Many years later, Whelan emphatically denied that this was the case; he said he took no position on the location at that time.

In any event, MacRae's expectations rose again during a meeting with Dr. Catherine Wallace, chair of the newly formed Maritime Provinces Higher Education Commission (MPHEC). The Commission was a child of the Council of Maritime Premiers, which had been set up in 1971 to promote regional co-operation. The MPHEC's role was to co-ordinate the evolution of post-secondary education in the Maritimes, and it eventually would have to decide whether there should be a fourth veterinary college, and, if so, where.

"Dr. Wallace said to me, 'You know it's pretty obvious that it should be in association with the school of agriculture,'" MacRae recalled. "She went back, and what happened then after the story got out and the deputy ministers all went back — UPEI got into the act and stated publicly that they wanted it there. Acadia got into the act and said they wanted it there. UNB got into the act and said they wanted to have it at UNB.

"At some time later she came to see me again and said, 'Well, we have a problem. Now everybody wants it. UPEI wants it. Acadia wants it. UNB wants it. So what are we going to do?'"

If UNB wanted the college, it didn't want it very badly. Dr. John Anderson, then president of UNB, was a biologist by training, with particular interest in fishing and aquaculture. He approved heartily of the idea of setting up a veterinary college in eastern Canada, but didn't think it necessarily should be at his university. He was new in the job; the university had financial problems, and a veterinary school, with its hospital, laboratory, and animal-care facilities, looked like an expensive, time-consuming proposition.

However, because he was asked — probably, he thought, by Dr. Wallace — to consider the college, he instructed Dr. Ron Taylor from his Biology department to explore the criteria for taking on the college, and whether UNB would be eligible. "He came back in about two weeks and said, 'Mr. President, I think we really should have this vet college.' I said, *'What?'* You had to have credibility in the other faculties, a lot of animals nearby, and so on. Anyway, we fit into all of them. I said, 'My goodness!'"

One university president who had no such ambivalence was Ronald Baker of UPEI. When word filtered to Charlottetown from Truro that the federal government was romancing NSAC, Dr. Baker sprang into action. He recalled that he went to Prince Edward Island Premier Alex Campbell, pointing out that the MPHEC, not a federal cabinet minister, should decide where the veterinary college would go. He also wrote a letter to that effect to Dr. Wallace. And in a letter to the Charlottetown *Guardian,* he suggested that UPEI would be the natural place for a veterinary college, and urged Islanders to "let their needs be known loud and clear — especially as they relate to veterinary science."[5] (This idea apparently caught the university Senate by surprise: according to the minutes of Nov. 14, 1974, Professor Andy Robb, referring to the letter, wanted to know at what point Senate would enter the debate as to whether such a college should be set up at UPEI. The president said the question would go ultimately to the MPHEC, and would become a matter for Senate if UPEI were considered a possible site.)

In a defining moment in the history of the Atlantic Veterinary College, the MPHEC did, in fact, seize the initiative, appointing Dr. Dennis Howell, then dean of the Ontario Veterinary College, to study the need for, and feasibility of, a veterinary school in the Atlantic region. Howell, accompanied by assistant Jean McDonald, spent the spring and early summer of 1975 touring the Atlantic provinces and meeting officials of educational institutions, various levels of government, and representatives of groups such as organized agriculture and veterinarians.

McDonald said later that one striking aspect of those meetings was how widely it was assumed throughout the Maritimes that the college would go to Nova Scotia. That assumption, she said, was particularly evident in Nova Scotia — so much so that the Nova Scotia government appeared to exert little effort to sell the province to Howell. "We had the impression that they assumed that they would get it without even trying."

In McDonald's estimation, Acadia University in Wolfville had a very strong case, noting, among other things, the university's strong academic base, its physical facilities, the proximity of the Agriculture Canada Research Station, and the high population of domestic animals near the university. McDonald was also impressed by the warm reception she and Howell received. "They were willing to co-operate in any way," she said.

The temperature at Dalhousie University was markedly cooler.

"It was quite obvious," McDonald said, "that Dalhousie had the same attitude as the Nova Scotia government: 'If you wish [the college] to be here, we will' — well, 'tolerate' is not quite the right word." In McDonald's opinion, the Dalhousie administration saw the proposed veterinary college as an institution that would not quite fit in with their existing professional schools.

Dalhousie submitted to the Howell commission a lukewarm report by a faculty committee, which observed that Dalhousie could provide the nucleus for a veterinary college with its Faculty of Medicine and strong arts and science programs, but lacked a department of animal science. Besides, the Halifax-Dartmouth area did not have the requisite livestock population. "When we were driving out of Halifax, we could see that there appeared to be no significant livestock population within twenty miles," McDonald said. "We said, 'They're sitting on a rock!'"

The University of New Brunswick dutifully submitted its bid, citing, among other points in its favour, the availability of livestock and small animals for clinical studies; enough space on campus for teaching and research; a sound base in the arts and sciences, as well as good academic services; and ready access to a number of fish and marine laboratories.[6]

Prince Edward Island's case, bolstered by supporting documents from farm and business groups, the veterinary profession, the city of Charlottetown, and the university, focused on the Island's strong agriculture base, as well as its burgeoning aquaculture industry. A high concentration of livestock near Charlottetown would supply a pool of clinical material, the nearest farm being no more than ten minutes away from the university. In addition, the provincial government promised to do "everything in its power" to accommodate a veterinary school, including acquiring property for a research facility, finding space on campus, and revising the University Act to enable the university to grant professional and graduate degrees.[7]

Aside from its inability to grant graduate degrees, UPEI's main handicap was that it lacked a nearby medical school. Ron Baker recalled discussing this issue with Howell when he arrived on the Island. "He was obviously sympathetic and thought that we had two strong arguments: the variety of animals on the Island and the fact that there was little danger of urbanization. But the strike against

University of Prince Edward Island campus in 1974

us was that we had no medical school. He said that no one in North America in recent years had set up a vet college without having a medical school nearby.

"When I told Mose Morgan, the president of Memorial University, about that, he said that Memorial had a medical school and that Eastern Provincial Airlines, the main carrier in the Atlantic provinces then, had direct flights between St. John's and Charlottetown. He would be our medical school." Morgan reiterated his offer in a letter to Howell, noting that Memorial had "no territorial ambitions" regarding the college.

In a sense, the Island's strongest case — although it was not one given high priority by the Howell commission — was simply that of fairness. At the time, the province had no professional schools and few degree programs. New Brunswick had the Maritime Forest Ranger School[13]; Nova Scotia had an embarrassment of riches — a law school, a medical school, a dental school, a school for the blind. "For all too long," the authors of the brief lamented, "Islanders have seen their sons and daughters forced to leave the province in order

Official opening of AVC, May 1987. *Left to right:* Jean McDonald, assistant to the dean, Ontario Veterinary College; Dean D. G. Howell, OVC, author of the Howell Report; and Dr. R. G. Thomson, Dean, AVC

to avail themselves of advanced education opportunities." In other words, it was payback time.

Many of the Nova Scotia presentations — including briefs from farm organizations, veterinarians, and municipal and county governments — plumped for the Truro site. They pointed to Truro's central location, the importance of the agricultural education component, the availability of animals, the existing infrastructure. The Agricultural College, banking on Dalhousie's support, talked about a split campus, with Dalhousie providing some of the advanced science work.[9] However, McDonald said that Dalhousie did not appear to be committed to assuming responsibility for the veterinary school as a faculty — at least, not committed enough for Howell's liking.

If hopes (or fears, depending on the point of view) of obtaining the veterinary college were rising at NSAC in the summer of 1975, it would not be surprising. As Howell was beginning his study, Principal MacRae recalled, he held a meeting in Truro with senior administrators of NSAC. "He stated clearly to all that the obvious and preferred location for a regional veterinary college was in Truro, associated with the Agricultural College and with easy accessibility to Dalhousie medical faculty and its library."

Coming at the outset of Howell's study, that statement seems extraordinary. Jean McDonald suggested later that it was a case of misinterpretation on MacRae's part — that Howell had indicated only that Truro would be *considered* as a site for the college — and it was. But Acadia outshone Truro, and would have been a stronger contender had the Nova Scotia government appeared more eager to get the college at that time. "We hadn't seen Acadia at that point," she said. "We hardly knew anything about Acadia then."

Dr. Ross Ainslie, a Halifax veterinarian, said he formed an impression similar to MacRae's at a meeting with Howell that summer. The Nova Scotia veterinarians pointed out that students attending a veterinary school in Truro could take some of their medical courses from Dalhousie. "He said, 'That sounds good. That should work. You can pretty well depend on it, that [the college] is going to come to this area." Jean McDonald's interpretation of Howell's comments was that he was referring to his belief that the other provinces would choose Nova Scotia.

In fact, Howell's decision ruled out Nova Scotia altogether. McDonald remembered flying back to Guelph with Howell, adding up the points in favour of each site under consideration. The result caused Howell some trepidation. "He said, 'What do you suppose Catherine Wallace will say?' We knew that our report was going to make waves throughout the area. The whole Atlantic region, when we first approached them, had seemed so certain what the result would be. Newfoundland had decided they were not going to be considered. PEI thought they didn't have much of a chance. And New Brunswick assumed it would go to Nova Scotia, as usual."

On August 31, 1975, Howell submitted his report to the MPHEC. His first choice was UPEI. UNB was next, followed by Acadia. In his report, Howell observed that UPEI had a sound and growing base in arts and sciences, and excellent library and computer facilities; a plentiful supply of animals nearby; adequate space on campus; an Agriculture Canada Research Station next door; and prospects of instituting graduate studies programs. In addition, the young university needed activities relating to the socioeconomic life of the Island.

Howell concluded that none of the top three institutions on his list had an overwhelming advantage that would make site selection obvious. But he added that UPEI would benefit greatly from the presence of a veterinary faculty. "The impact of the school on the Province as a whole would be profound, more so than that of virtually any other professional school."[10]

When Howell's recommendation was announced, President Anderson of UNB heaved a sigh of relief. "I was delighted," he said, "not so much that we didn't have to worry about having to set up a vet college, but I felt that this would be great for UPEI, great for the Maritimes. I still feel that way. UPEI was a small university that needed its academic place in the sun, and it wasn't going anywhere the way it was, however good it might have been." Catherine Wallace, according to Jean McDonald, was also delighted. "She said, 'This is what the Maritime provinces need. They need things that will pull them together, not pull them apart.'"

Predictably, administrators at the Truro college were not so pleased. In Howell's report, Truro did not even make the short list. He cautioned that "very careful thought" would have to be given before Truro could be considered as a possible location. He conceded that veterinary schools traditionally had been located on a campus with an agriculture school. "Truro is, however, not the site of one of Nova Scotia's universities and veterinary medicine must be considered as a university faculty."[11]

Officials at the Agricultural College were "appalled" by the report, MacRae said. "We couldn't find any logic in why he rated UPEI first and put us in fifth place. And his argument, in spite of all the things we tried to tell him, was, 'Well, you're not a university, and you don't have a library, you don't have this, don't have that — which was an excuse, really. Because we were well on the road to the degree thing. The groundwork had all been done, and Dalhousie was very supportive."

MacRae was still angry twenty-seven years later. "It was the process that really made me angry," he said. "For someone to sit in front of you and say, 'This is where it should be, and this is what I'll recommend, and then turn around and come to a totally opposite conclusion — it was a little bit hard."

If Howell did change his mind, what swayed him? Former Premier Alex Campbell said the Island government's strong presentation carried the day. "It is my understanding that the submission made by PEI to the Howell study was critical in establishing the merits of a PEI location," he said.

A letter to Campbell from Howell in November 1976 lends weight to that theory. Howell wrote: "The recommendation that the proposed vet school should be in Charlottetown became increasingly clear as I progressed with the study, having regard to the criteria which were defined at the beginning.

"I must confess, however, that I was to some extent motivated by the enthusiasm which we received from all segments of opinion in the province, and particularly from your Government.

"A school of this kind requires strong government support if it is to succeed and I have no doubt in my mind that with the support which I am sure will be forthcoming the Charlottetown school will in fairly short order become a prestigious institution on a regional, national and international basis."[12]

Jean McDonald said it was that supportive attitude, plus the variety of animals in the area and the proximity of research laboratories, that impressed Howell most.

Almost three decades later, however, Herb MacRae remained convinced that the decision was a political one — that Howell had been influenced by federal politicians. "It is in many ways a classic tale of how decisions are made on political grounds," MacRae said, "and who your friends are, rather than on academic or any other practical grounds."

Jean McDonald insisted that Howell was not influenced by any politician. It was a project in which facts, not politics, were the determining factor, she said. Whelan said there was no political interference on his part: he had remained neutral on the issue of the site until the Howell report came in.

If politics was not a factor then, it certainly became one in years to come. Howell's report did not settle much of anything. Whether Atlantic Canadians knew it or not, a political poker game that would last almost a decade was just beginning. New Brunswick, which supported the Island as a site for the college, and Newfoundland, which offered no resistance to the Island site, were not major players in the game. But governments in Ottawa, Halifax, and Charlottetown would now indulge in years of bargaining, threatening, bluffing, and posturing. The federal government's position was that all four Atlantic provinces had to agree on the location of the veterinary college before Ottawa would contribute its half of the construction costs. Nova Scotia balked. Its Liberal government, led by Premier Gerald Regan (whose Minister of Agriculture, Jack Hawkins, was a former teacher at the Agricultural College), insisted that if a college were built, it should be built in Truro. The issue surfaced repeatedly at meetings of the Council of Maritime Premiers. The years went by. Nothing was settled.

In January 1977, Premier Alex Campbell wrote to Whelan, proposing two solutions to the impasse: would Ottawa fund construction of the college without Nova Scotia's participation? And might the Canadian International Development Agency (CIDA) contribute in Nova Scotia's place? CIDA's involvement would mean that students from developing countries could attend the college. Aid to developing countries was a subject dear to Whelan's heart, and the CIDA solution was one that was to find its way into the eventual agreement on the college. To Premier Campbell's query about CIDA, Whelan replied that Campbell should contact the agency president directly. On the other point, Whelan's reply is surprising, given his consistent position later on. "The approval in principle for my Department to provide capital assistance for the establishment of a Veterinary College in the Atlantic Region is not contingent upon the location or involvement of all four Atlantic Provinces," he wrote. "While it would be desirable for all four provinces to be committed to such a project, it would certainly be possible to proceed with only three provinces being involved."[13]

Later, Whelan did an about-face, pointing out that Ottawa's contribution would be limited to capital costs, and, for ongoing financing, all four provinces needed to be on board. There was no sign of that happening by the end of 1977. In December, the issue was not even on the premiers' agenda at their formal meetings. The Island and Nova Scotia governments were firmly entrenched in their positions, and apparently felt there was no point in rehashing them.

It was just as well. Some members of the public were getting tired of hearing about the interprovincial squabbling. In December 1977, Rev. A. S. Rockwell, pastor of Grace Baptist Church in Charlottetown, used the dispute as a launchpad for a mini-sermon during a religious broadcast on radio station CFCY, Charlottetown:

> "Having been born in Nova Scotia myself," he said, "I am ashamed of my native province's attitude in this instance, if the newspaper reports are correct. It all smacks of poor losers and bad sportsmanship. They are trying to change the rules in the middle of the game.
>
> "But isn't that just like many of us, i.e., we go along with God as long as He does nice things for us. But then when He speaks of the unpleasant facts, we welch, and try to cop out. Sin isn't pleasant, but it is a fact. Judgement Day is not a happy prospect, but it must be faced, and prepared for now. Just to say that we disagree with the Bible because it doesn't suit us, is as wrong as Nova Scotia saying, 'We don't agree with the commission's findings, so we bail out.' Only with God, you don't bail out...."

The University of Prince Edward Island was not prepared to bail out. In December 1977, President Baker told an *ad hoc* college planning committee that he had decided to advertise for a "Senior Administrative Officer" for the college. The term "Dean" could not be used, he explained, until the university Senate had created a new Faculty of Veterinary Medicine. "He [Baker] emphasized that the University was proceeding positively despite the static being generated by Nova Scotia," the December 23, 1977, committee minutes reported.

Supporting those positive steps was the government of Prince Edward Island, which had allocated $200,000 toward the planning of the college. By 1978, the Island government was preparing to go even further. In an attempt to force Nova Scotia's hand, Agriculture Minister Bud Ings announced in August that the Island government planned to go ahead with the veterinary college, with or without the other provinces.

It was election year in Nova Scotia. Ings said later that Regan wasn't necessarily personally opposed to the Island site, but was afraid of losing seats in the Truro area over the issue. "Regan's attitude was, 'You fellows go ahead and we'll come in after the election and go along with you. But right now we can't announce it because the bunch around Truro are so adamant about getting the new vet college to go to Truro.' So we more or less developed the attitude

Dr. Peter Meincke, second president of the
University of Prince Edward Island

that we'd go ahead. We were trying to convince everybody that we were serious."

Unfortunately for the Island government, Whelan wouldn't play its game of bluff. "Whelan had to back off," Ings said, "because he didn't want to hurt Regan and the local Liberals, saying, 'We're going to build it anyway.' It would kind of make a fool out of Regan to go behind his back. So it was a combination of a lot of things that set the project back."

In September, Regan lost the election anyway. A Conservative government, under Premier John Buchanan, took over.

That same month, the new president of UPEI, Dr. Peter Meincke, announced that the university had found the college administrative officer it had been seeking. From then on, the new official was called the planning co-ordinator, and his duties included determining "the feasibility and appropriate orientation" of a school of veterinary medicine. As it turned out, one of his most important functions was to sell the project in the Atlantic region. If and when the college was built, he would be offered the position of dean.

Howell and McDonald had already sold the almost-dean to the University of PEI. McDonald recalled, "We said, 'Don't worry about it. We have the perfect person for you.'" That "perfect person" was Dr. Reginald G. Thomson, then head of the Pathology department at the Ontario Veterinary College. The new dean-in-waiting was to start work in January. From Day One, and for more than a decade to come, he would have a profound impact on the University of Prince Edward Island.

3: The Tug of War Begins

In many ways, Reg Thomson was perfectly suited to his new job in Prince Edward Island. He was a country boy who had grown up on a dairy farm three miles outside of Woodstock, Ontario. In high school when he started dating Helen Ure, a town girl and the love of his life, he'd call for her in a green pickup truck with a picture of a big Jersey cow on the side. When he graduated from the Ontario Veterinary College, he worked in a large-animal practice in Ontario for two years. He was no stranger to hard physical labour, and had no trouble communicating with farmers, or understanding their problems. In his rare moments of spare time, he liked collecting antique farm implements and photographing old barns.

But Thomson was also a brilliant pathologist who eventually obtained a doctorate from Cornell University, returned to Guelph to teach, rose quickly through the ranks to become a full professor and a department chair, edited *The Canadian Journal of Comparative Medicine* for fourteen years, and wrote numerous papers and two textbooks, one of which was translated into several languages. In short, he had credibility in the academic world.

He also had the kind of drive and determination needed to push the derailed veterinary college back on track. Dr. Bob Curtis, who graduated from OVC two years after Thomson and eventually became a faculty colleague, recalled that, even as a student, Reg was determined to win, whether it was at football or in student politics. He usually did. One rare exception was his bid for a seat on the student council at OVC, his competition being the candidate from the much larger agriculture school. He didn't succeed. "In all the time I knew him," Curtis said, "that was the only thing he went after that he never got." But it wasn't for lack of trying.

Quiet, intense, and extraordinarily well-focused, Reg Thomson lived and breathed veterinary medicine, particularly his great love, pathology. When he was teaching and writing books at Guelph, Cur-

Dr. Reg Thomson

tis said, he'd work at the university until about five o'clock, go home for supper and an hour's nap, and then return to the school until midnight. "He would do that six, seven days a week, for years. It was unbelievable how hard he worked."

In the fall of 1978, at the age of forty-five, Thomson took a two-year leave of absence from the Ontario Veterinary College, moved to Charlottetown with his wife, Helen, and one of their three daughters (the other two were in university), bought a house on Prince Charles Drive, and in January set up an office on campus, first in one of the university residences, and later in a renovated pig barn.

His job was to persuade Atlantic Canadians of the need for, and benefits of, a veterinary school in the region, and to come up with a vision of the kind of school it should be. He tackled this task with the kind of single-minded devotion he had applied to every other cause in his life. In planning the new college, he wrote hundreds of letters to universities, bureaucrats, politicians, veterinarians — anyone who could suggest how a veterinary college should be run. Armed with flip charts, he travelled throughout the Atlantic region, sometimes with his family, sometimes with UPEI President Peter Meincke or staff from the premier's office, talking to the media, farmers, veterinarians, bureaucrats, women's groups, business people, politicians

— anyone who would listen.

"He was very eager for the construction to start, and he wanted to be the first dean," Curtis said. "He touched all the bases. Somebody once said that if you called him [to a meeting] and said you weren't sure anybody was going to be there, he'd come anyway — just in case two people showed up."

Prince Edward Island had a new government by this time — Conservative Angus MacLean took over as premier in 1979 — but the province was still keen on obtaining the veterinary school. David Weale, who was principal secretary to MacLean, said Thomson could always be counted on to work for the cause. "He wasn't a flashy man by any means, but, my goodness, he was committed to the thing happening. Persistent? Yes. Relentlessly persistent. Indefatigably persistent. To the point where I felt worn down by the man at times."

Thomson told his audiences about the importance of preventive medicine, not only to farmers but also to consumers. He told them how essential disease control and prevention were in the growing aquaculture industry, and how a local veterinary college could contribute to that industry. He talked about the importance of research. He talked about economic spin-offs from a veterinary college. He talked about the annual influx into Canada of foreign-trained veterinarians and the lack of educational opportunities for young Atlantic Canadians.

"Your presentation was very well done," Barbara Doughart of the provincial board of the PEI Women's Institute wrote in a thank you note to Thomson after a Monday morning speech in January 1980, "and left no doubt in the minds of any of us as to the validity of such a proposal, even though some of us may have had reservations in the beginning."

Others were not so easily persuaded. Even the university community on the Island was divided on the issue of the veterinary school. Glenn Palmer, who at one time was chair of the Chemistry department, said people in the Science faculty were supportive; they saw the veterinary college as a means of stabilizing enrolments and bolstering the university's reputation. But some members of the Arts faculty were opposed to any kind of professional school. "This was going to be a liberal arts college, and anything professional would sully that reputation."

Even David Weale, who had taught history at UPEI before obtaining a leave of absence to work for MacLean, could not muster much enthusiasm for the proposed school. He therefore was in the uncomfortable position of having to strive for a prize he wasn't terribly keen on winning. "It was kind of naive and idealistic, I suppose," he said, "but I could see there was a place for a liberal arts education that didn't have all those professional schools attached to it."

President Meincke, a physicist from the University of Toronto who had arrived shortly before Reg Thomson, recalled that he was at first neutral about the college proposal. He became a strong supporter as he learned the benefits the college might bring. One of the arguments that convinced him was the net outflow of dollars for Island students studying in Nova Scotia in fields such as law and medicine. The Island government was paying the government share of tuition for these students, without receiving anything comparable from the other Maritime provinces. "To me, this was the only way in which we were going to get a balance of payments," he said. "There wasn't anything else that would make sense. At least, I didn't see anything on the horizon. So the flow of money from Nova Scotia to pay for the veterinary college would be balanced by the money going from here to pay for the medical school."

The financing of higher education in the Maritimes was a perennial sore spot for the Island. In the late 1970s, the Council of Maritime Premiers had agreed to a regional funding formula for higher education. It meant that if, say, a student from the Island went to Nova Scotia to get a degree in a program not available at home, the Island government would pay the government share of the cost of his or her education. Similarly, if a Nova Scotia student wanted a forestry degree from the University of New Brunswick, Nova Scotia would pay the government share. On a per capita basis, the Island had the biggest outflow of money — more than $3 million by the early 1980s — because it had the fewest degree programs and no professional schools.[1] New Brunswick had a Faculty of Forestry and a Faculty of Engineering at UNB, and was negotiating with the other two provinces for cost-sharing of new facilities for the Maritime Forest Ranger School; Nova Scotia had an embarrassment of riches — a law school, a medical school, a dentistry school, a school for the blind. Emery Fanjoy, formerly secretary to the Council of Maritime Premiers and, later, to the Conference of Atlantic Premiers, said the funding formula was weakening the economy of the Island and the fabric of its new university, and both had to be strengthened in some fundamental way. "The only thing that they had that was really important on the horizon was this veterinary program. It involved the other provinces, which meant a flow of students and therefore money from the other provinces into UPEI."

That was not an argument that cut much ice in Nova Scotia, although Premier John Buchanan was more sympathetic to the Island's aspirations than were some of his colleagues. There were, as Buchanan delicately phrases it, "some classic discussions" in cabinet and caucus over the issue. The problem was that eight of the Conservative MLAs were from ridings that would be affected directly by a new development in central Nova Scotia. Henry Phillips, one of the Island government's negotiators, recalled that one of the strongest opponents was Ron Giffin, the Finance Minister, who was from Truro. "I remember him telling me once at a cocktail party that the college would never come to PEI," Phillips said. "I said, 'You can never be too sure about those things.' Whether it was just because of NSAC or because he was Finance Minister and it was just the money, I don't know. I got the sense that it might have been both."

Certainly, Nova Scotia would be paying a large share of operating costs. And wherever the veterinary college went, it would piggyback hundreds of new jobs. Not surprisingly, the Nova Scotia farm community was strongly opposed to letting the college go. The agriculture minister, Dr. Gerald Sheehy, a veterinarian, even maintained that a fourth veterinary college was not needed at all.

Buchanan, who served as premier for twelve years before resigning in 1991 to take a seat in the Senate, said the main issue was money. "Being the largest province involved, we would have to put up a lot of money, capital funds for construction, and a percentage of the administrative costs. As I recall, the majority of that was going to be paid by PEI,[2] and the federal government would be providing money, too. Still, we would be second, and New Brunswick third. So that was a problem for us. It was a question mark, whether the taxpayers of Nova Scotia should put that much money into a veterinary college in Prince Edward Island.

"The second concern was that we would be funding a college in PEI rather than in Nova Scotia, when there was a school of thought that the appropriate place for a veterinary college would be in Truro, where the Nova Scotia Agricultural College was."

At one point, Thomson and Meincke paid a visit to the heart of opposition, the Nova Scotia Agricultural College. "It was really an eye opener," Meincke recalled, "because we discovered that they were absolutely convinced that if the veterinary college went anywhere else but Truro, the Agricultural College would be moved somewhere else. No wonder they were fighting so hard to get it, and Truro is a very strong political centre of Nova Scotia, so it made it very hard for the Nova Scotia government at the time.

"We weren't looking to [remove the Agricultural College], but anything we said didn't make any difference to them. We couldn't convince them it wouldn't happen. So that was an eye opener, but it didn't help us at all. We just kept working away at it."

Farm organizations in Nova Scotia continued to lobby Liberal MPs, the Buchanan government, and the federal government for the Truro site. "We have carefully studied the Dr. Howell report," Rolland Hayman of the Nova Scotia Institute of Agrologists wrote to Whelan, "and cannot accept either the report or the recommendations."[3]

Some veterinarians questioned whether there should be a fourth college at all. Dr. Ross Mitton, a retired veterinarian in River John, Nova Scotia, submitted a number of articles to the local newspapers, arguing that there already was a surplus of veterinarians, that the college would prove to be a "disastrous" financial burden, and that the money would be better spent on bolstering the Ontario Veterinary College in Guelph. "We must ask: Do we really need it?" he wrote in the Halifax *Chronicle-Herald*. "And can we not do without it? Many of us believe we do not need it; in fact, many of us believe another veterinary college in Canada could do more harm than good."[4]

The Canadian Veterinary Medical Association (CVMA) also contended that a fourth college was not needed, at least at that time. Citing high capital and operating costs, the scarcity of faculty, and the uncertainty about manpower needs, CVMA president Dr. D. L. T. Smith wrote in a letter to Island MP Tom McMillan: "The CVMA and others have conducted detailed manpower studies, and these are updated annually. None of these studies as yet have demonstrated an unequivocal need for more veterinary graduates on the North American continent. In fact, the reverse tends to be true."[5]

That opinion seems at odds with a manpower report conducted for the CVMA in 1977. The authors observed: "Most veterinary manpower studies have suffered from being too conservative. In 1976, the most optimistic projections of the Jones-Moynihan report [issued in 1971] have been substantially exceeded, and even though an agricultural recession has existed for several years, employment opportunities for veterinarians still seem not to be met."[6]

Dennis Howell, former dean of OVC, concurred with that opinion. In a letter to Whelan in June 1981, he wrote: "You will wish to know that the feared surplus of veterinarians voiced by certain elements in my profession has to date proved inaccurate." The colleges at Guelph and Saskatoon were turning out about 175 graduates a year, and "as many as four times the number of graduates emerging from Guelph and Saskatoon this year could have found jobs."[7]

Nevertheless, the Ontario Veterinary Association also mounted a campaign against the proposed new college. In a letter to Whelan, association president Dr. Clayton MacKay said the three existing schools were seriously underfunded, to the point where the Ontario and Quebec schools were in danger of losing accreditation. MacKay questioned whether a fourth school would have sufficient operating funds, and whether the existing schools would be "pirated" for faculty. Moreover, he said, past veterinary manpower reports in Canada were simply guesstimates. "In the face of this present situation, it appears ludicrous to us that the federal Government should push on with a commitment of many millions of dollars to yet a fourth veterinary school in the Maritimes, where the real need has simply not been established or proven."[8]

Whelan wanted a fourth college, and he wanted it in Prince Edward Island. The Ontario veterinarians were not pleased. "I remember a big meeting in western Ontario, where the veterinarians attacked me for going ahead with the school," Whelan said. "They said we didn't need more veterinarians. See, we had worked on an expansion for programs of veterinary medicine in Saskatoon and we made a big addition at Guelph and to the one in Saint-Hyacinthe. I gave them hell. I said, 'Just because you educate somebody as a veterinarian doesn't mean he is necessarily going to be a veterinarian.' But they resisted until the sod was turned."

By the winter of 1979–80, three different premiers had taken over the reins of the Island government since the Howell report chose Charlottetown as the preferred site for the college. The federal government had also changed hands — Joe Clark's Conservatives were in power, briefly — as had the governments of Nova Scotia and New Brunswick. What hadn't changed was the political manoeuvring, with its many peaks and valleys, over the veterinary college. Reg Thomson kept soldiering on. "I used to call him Don Quixote," Helen Thomson said. "I'd say, 'I don't know how you can keep doing this.'"

One period in which the talks must have been going well was at the beginning of 1980. Thomson had been promoting the veterinary college on a radio program, and he received this letter from Moncton, handprinted in neat, elementary-school style:

Dear Doctor thompson.
I am 8 years old I am in grade 3
My name is Geoffrey Johnston
I want to Be a veteU.P.EHarian
Wen I am old enough.
I like animals there is a shortage
of gasoline, in that matter, We Will
need more animals for transportation
and for food, but there are 2
provinces that Want the vetenarion
college those 2 provinces are
pricE Eward island and Nova scotia
I Thingk it shoud be Nova seotia
Because Nova scotia Has the
agriculture college and Nova scotia
ould use it more because Nova scotia
Has the agriculture college and
the people that woud Work
at the agriculture college
could go there to learn more about
animals I herd you on the radio
and it was very interisting
these are, my ideas Not My Mothers or
fathers. idea.
RR4
moncton, NB
EIC 8J8
Geoffrey Johnston

Thomson replied:

January 28, 1980

Mr. Geoffrey Johnston
R. R. #4
Moncton, N. B.
E1C 8J8

Dear Geoffrey:

        Thank you very much for your letter. I am very pleased
to hear that you are interested in Veterinary Medicine, and also
that you would like to see a new college in the Atlantic region.
I hope that will be a decision on the location of the College
before long, and perhaps you could watch the papers over the next
month or so and see if you see something about where it is going to
be. If you are in Charlottetown some time please look me up, as I
would be pleased to discuss the subject further with you.

                        Sincerely,

                        R. G. Thomson
                        Planning Coordinator

RGT/m

The premiers did not settle matters in "the next month or so," but that summer it looked as though an agreement might be close at hand. The minutes of the Council of Maritime Premiers meeting in June noted that the Atlantic premiers were "close to agreement that Charlottetown would be the site of the proposed college." The main outstanding issues were the assurance of federal financial support and the development of a suitable plan to integrate the programs of the veterinary college and the Agricultural College in Truro.[9]

By the time the premiers met in September, they still had not wrestled with the issue of the Agricultural College.[10] And Thomson still had his worries. That fall, he wrote a memo to David Weale, warning that if the provinces couldn't agree soon, Ottawa might switch its support for a fourth school to British Columbia. "The premiers should be reminded of this point in case they think the College will be kept on ice until they wish to proceed," Thomson wrote. "The rumours from British Columbia are not just idle comments. One of the strong factors for B.C. would be the value of the Fish Health component."

Thomson also maintained that the Ontario Veterinary Association, which opposed a fourth college, was manipulating the Ontario Veterinary College behind the scenes. The OVC, he said, had stopped adhering to its quota of about six Atlantic students a year. "It is important to realize," Thomson wrote, "that their action has been quite deliberate, particularly this year when the highest number of Atlantic students ever were admitted. The new dean at OVC allowed himself to be influenced by the Ontario Veterinary Association, which does not see a new school here to be in its best interests. If the college here does not start, the enrolment from the Maritimes to OVC could decline and Ontario would once again have controlled a situation to its advantage but to the disadvantage of the Atlantic region. These statements would never be acknowledged as true by OVC or the OVA, but, in fact, that is what happened. The Minister of Education is aware of the situation but is reluctant to intervene and run the risk of being accused of intervention into a political problem in other provinces."[11]

Premier MacLean shared Thomson's foreboding about the stalled talks. In a letter to Premier Buchanan in December 1980, MacLean warned that further delays might prompt Ottawa to direct the proposed college to British Columbia. "Even if this were not to happen, a further delay in these already protracted negotiations could very well have the effect of dooming the project to a slow death by delay and indecision." MacLean added that a failure to reach an agreement would place a strain on working relations among the Maritime provinces at a number of levels, and "would have the effect, in the minds of many Maritimers, of seriously eroding the foundations of Maritime co-operation and good will which has built up over the years."[12]

In the New Year, Thomson's natural optimism surfaced in a letter to Dr. W. M. Adams, associate dean of the veterinary school at North Carolina State University. It ended this way: "We continue to be optimistic although certainly a good deal of time has gone by. We are dealing with parochial politics and not rationality. We should have good news in the near future."[13]

The Prince Edward Island government apparently decided to create that good news. In February 1981, Premier MacLean announced in the Legislature that the province had decided to proceed with the veterinary college, with or without Nova Scotia. Construction could start within a year. Because the Maritime governments had been unable to agree on a location, he said, the region was falling behind the rest of Canada in veterinary services and research. "We cannot tolerate any further delay," he said. "This service is required to fully utilize our agricultural and fisheries industries. For these reasons, we have decided to act, and act now."

MacLean said the government was taking this action with the support of the governments of Newfoundland and New Brunswick. Ottawa had indicated it would contribute capital funds, and planning for construction would start once cost-sharing arrangements with Ottawa were finalized. He added, "We feel confident ... that Nova Scotia will see fit to participate in our vet college, as we participate in their medical, dental, law, engineering and agricultural schools."[14]

MacLean, frustrated by lack of progress on the college issue, apparently had hauled out the previous Liberal government's strategy of trying to force Nova Scotia's hand. He proposed that Prince Edward Island would provide interim funding for Nova Scotia's share of the construction costs, with the expectation that Nova Scotia would contribute its share of operating costs through payments made for students attending the school. He based this expectation on Buchanan's promise that Nova Scotia would ensure that its residents could attend the veterinary college if it were built in Prince Edward Island.[15]

According to the University of Guelph's Dennis Howell, who waded into the debate in June, Catherine Wallace, Chair of the Maritime Provinces Higher Education Commission, fully supported MacLean's proposal for getting the college project moving. In a letter to Whelan, Howell observed that MacLean's proposal for covering capital costs was the same as the funding agreement that enabled Newfoundland students to attend dental school at Dalhousie University. "Dr. Wallace considers this a satisfactory arrangement," Howell wrote, "and will work out (as her office did in the dentistry situation) the precise amount of this figure...."[16]

As Peter Meincke pointed out in a letter to Whelan in July, MacLean, in fact, had considerable support in the Atlantic region for the Charlottetown site. Besides the MPHEC, supporters included the Atlantic Provinces Economic Council, the Atlantic Provinces Chamber of Commerce, the governments of New Brunswick and Newfoundland, and some of the region's universities. "We believe that the above evidence constitutes a sufficient degree of support for the Federal Government to authorize the construction of the College," Meincke wrote. Meincke also observed that the Prince Edward Island government was transferring $1.8 million a year to Nova Scotia for Island students enrolled in programs not available at UPEI. "This is very close to the amount Nova Scotia would transfer to Prince Edward Island to support the operation of the veterinary college," he said. "The College is the only forseeable facility that will balance such educational payments." Surely, Meincke argued, some matters were important enough to go ahead without the full public agreement of all the provinces. "Surely the Federal Government would not deny the facility to the region simply because Nova Scotia appears to be prepared to see it built in British Columbia rather than lend its full public support to Prince Edward Island."

None of these arguments managed to sway Ottawa. Unfortunately, it was an inauspicious time to ask any government for money. By the beginning of the 1980s, the region, along with the rest of North America, was heading into the worst recession in thirty years.

"At that time, we were going into deficit financing," Buchanan recalled. "Back in those days, when you had to ensure that the social net was looked after, that is, education, health and social services, the last thing on our minds was to put out a big chunk of money in capital funds and administration costs for a veterinary college in PEI."

The ups and downs of talks over the college were hard on Helen Thomson. She had no idea from one week to the next whether the family would be settling on the Island or moving on. "The problem I had personally was that our life was not being controlled by us anymore," she said. Her husband was still officially on leave from the University of Guelph, and he had to decide by September whether to return to Guelph or look for another job. "His leaving would be a disastrous setback for the project," Meincke told Whelan, "not only for UPEI, but for the region. However, I cannot in good conscience ask him to devote any more of his career to a project which is subjected to so much unnecessary delay."[17]

Even Reg Thomson, the single-minded optimist, was becoming discouraged. To cheer himself up, he took a course at the University in religious studies and worked on his second book. Then he decided he had done all he could do in Prince Edward Island. That fall, he accepted a job at the Western College of Veterinary Medicine in Saskatoon. The family sold the house in Charlottetown, said good-bye to Maritime friends, and headed west to start a new life.

4: Try, Try Again

As Reg Thomson was leaving the scene in Prince Edward Island, a new premier was arriving. Jim Lee, a mild-mannered realtor, succeeded Angus MacLean in November 1981, and immediately seized the veterinary-college issue by the scruff of the neck. Jim Lee wanted that college so badly, he earmarked it as his top priority during the first years of his mandate. As he saw it, the University of Prince Edward Island was just getting its feet planted in the ground and needed sustenance; the project would bring hundreds of temporary and permanent jobs to the Island; and the college could become the nucleus of all kinds of satellite industries related to veterinary work. Lee thought it was urgent to get the college issue settled once and for all.

"It had been dragging on since 1970," he said later. "We were then well into a decade or more with nothing concrete happening, just a lot of talk. It had to be now because this was starting to get stale. People were backing away from it and new players were coming on the scene. We felt it had to happen in the early years of our term because if we couldn't see it happening we were prepared to abandon it and move on to other things. It was going to be 'do or die.'"

Underneath Lee's bland, Clark Kent exterior, there beat the heart of a man of action. When the developer who was building Charlottetown's largest hotel ran into financial trouble in 1982, Lee's government took charge of the half-finished project. One of his first steps was to call a meeting of bankers and federal and provincial officials. On a Sunday afternoon, as the meeting began in the fifth-floor Executive Council offices in Charlottetown, Lee informed the group that nobody, but nobody, was to leave the building until an agreement was reached to refinance the project. About twelve hours later, the deal finally was consummated, and the hotel construction went ahead.[1]

"I guess I come from a background of how business people think

WHOSE MOVE ?

NS Premier John Buchanan playing Vet College Chess with PEI Premier Jim Lee (*The Evening Patriot*, Februrary 27, 1982), p. 4. Used with the permission of the artist, Dean Johnstone.

and operate," Lee said. "If you have a problem, sit down and discuss it. If you can work it out, fine. If not, move on to something else."

Also among Lee's strengths were his good connections in all four Atlantic provinces. All four provincial governments were Conservative. Lee had worked behind the scenes for years as a party organizer, and he had formed a friendship with Nova Scotia's John Buchanan a dozen years earlier. From then on, their families saw each other socially. Richard Hatfield of New Brunswick was not only a friend but also a supporter in the veterinary school tug-of-war. Lee had strengthened ties with Newfoundland's Brian Peckford through an exchange of trade missions with that province in the summer of 1982.

The three Maritime premiers had had many private discussions about the college, and Lee was pretty sure that Buchanan would come around eventually. "Premier Buchanan could see that Nova Scotia was doing very well with the number of colleges and other educational institutions they had," he said, "and Premier Hatfield and I made that argument to him many times: 'We have to share.' I think he was buying time to bring his colleagues onside, and I think deep down John Buchanan was supporting PEI."

Buchanan, of course, continued to have his own troubles. It wasn't easy to justify sending dollars to the Island during a time of fiscal restraint, particularly when many members of his own caucus viewed the college as a potential boon for their constituencies. Fortunately for the Island government, the federal agriculture minister, Eugene Whelan, emerged as a strong ally. He believed firmly that a fourth college was needed, and that it was needed in Prince Edward Island. His interest in veterinary medicine stemmed from his youth on a southern Ontario farm, when he and his brother Tom, both big and strong, used to help the local veterinarian with some of his more distasteful tasks on neighbouring farms. "They used to call us 'the mean Whelan boys' because of some things you had to do — for instance, when a cow was going to have a calf and she didn't develop enough, you had to cut the calf up inside. You saved the cow, and that was the most important thing for the farmer."

Whelan also was, in his words, "an eager-beaver nut" for research, especially aquatic research. There was a big fish-processing industry in his riding, and, in his early years in Parliament, he had travelled extensively throughout Atlantic Canada as Parliamentary Secretary to the fisheries minister of the day. That turned him on to the potential for fish research, and he was determined that a new veterinary college would include a fish-health component. By the early 1980s, he was already referring to the proposed school as the "veterinary and aquamarine college."[2]

Whelan had a number of reasons for supporting Prince Edward Island. For one thing, the Island was the first choice of the 1975 Howell Commission. Besides, Whelan happened to have a soft spot for the province and its farming population — a sentiment nurtured through his many visits to the Maritimes as agriculture minister. "I called it the Garden of Eden. What they used to say down there every time they got into trouble was, 'We'll get Uncle Gene to help us.'"

Most important, perhaps, was Whelan's friendship with Daniel J. MacDonald, a one-legged, one-armed farmer-politician from Bothwell, Prince Edward Island. "Danny Dan," as he was known on his

Honourable Dan MacDonald, Minister of
Veterans Affairs in the 1970s. MacDonald
promoted the veterinary college in cabinet.

home turf, had lost an arm and a leg to an exploding shell in Italy
during the Second World War. After the war, he returned home to
his farm, dug a cellar by hand with the help of his brothers, built a
house and barn, and started running his mixed farm by himself. He
milked cows by hand, carried buckets of water with a hook attached
to his left shoulder, used his wooden leg as a chopping block for
cutting potato sets, and operated tractors and combines by himself.
"There was nothing he could not do," said his widow, Pauline Mac-
Donald. "He was very determined. He would never, ever give in."

Not surprisingly, when he ran for the Liberals in the 1962 provincial election, he won a seat in the Legislature and kept it for the
next decade. He served as Minister of Agriculture and Forestry until
1972, when he ran, successfully, for Parliament, and immediately
caught the eye of Prime Minister Pierre Elliott Trudeau, who appointed him Minister of Veterans Affairs. Trudeau admired Dan
MacDonald's grit, resourcefulness, and self-reliance.

"Trudeau thought the world of Dan MacDonald," Whelan recalled. "Dan never said that much in cabinet. It had to be something important before he'd take the time of the cabinet to get his point across. He'd put his one hand on the table and say, 'Mr. Prime Minister,' and everybody in the cabinet knew that when Dan talked, the Prime Minister wanted to hear what he was going to say. Everybody would be as quiet as could be."

Whelan and MacDonald sat next to each other at the cabinet table. Their families became friends, and visited back and forth. (MacDonald once took Whelan out to the barn so that his pet bull, Fulton, could meet the federal Minister of Agriculture.) "We used to talk a lot about agriculture," Whelan said. "We got along really well because he understood agriculture. He used to say, 'Only a stubborn SOB of an Irishman could be Minister of Agriculture and take all that guff.' That job can be lonely, especially as the country becomes more urbanized. So it was a welcome addition to cabinet when Dan MacDonald arrived. We spoke the same language."

One of the subjects the farm boys discussed from time to time was the need for a fourth school of veterinary medicine in the Atlantic region. Whelan said MacDonald started badgering him shortly after arriving in Ottawa. "He'd say, 'Hey, Whelan boy, don't forget the greatest agricultural province in Canada.' He was a promoter of everything for PEI, of course, but a strong promoter of the school of veterinary medicine. When it was delayed he was very concerned — you could go so far as to say he was upset about it."

At the time, MacDonald happened to have a close link to the University of Prince Edward Island: his son-in-law, Dr. Kenneth Grant, was chair of the board of governors, and no doubt MacDonald was well aware of concerns that the higher-education formula was weakening the fabric of the University. The Island, with few degree programs and no professional schools, was sending an inordinate amount of money to the mainland to pay for higher education. In the opinion of Emery Fanjoy, formerly secretary to the Council of Maritime Premiers, the other Atlantic provinces, including Nova Scotia, recognized that inequity, and that is why they eventually came to an agreement. But, before that happened, the twists and turns of the political dance continued. First, the Liberal government in Ottawa blamed the Conservative Atlantic premiers for the delay. Then the

Some of the products used by nineteenth-century farmers and veterinarians

Basher, mascot at the Veterinary Teaching Hospital (1993)

Aerial view of AVC under construction during the winter of 1985–86

1989 Toxicology group (*l-r*): Elizabeth Holmes, Brian Grimmelt, Orysia Dawydiak, Steve Gould, Jennifer Brown, Vince Adams, and Dr. Peter Nijjar

Dr. Reg and Helen Thomson in 1994, a few years after his departure from the college

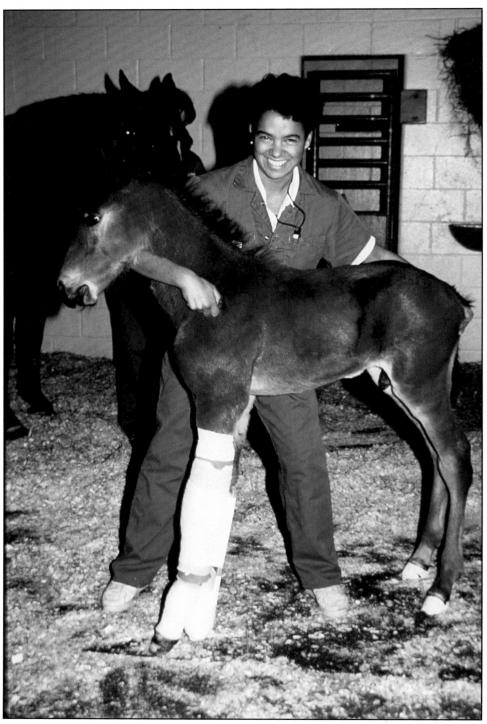

Claudia Campbell of the Class of 1992 with patched-up foal

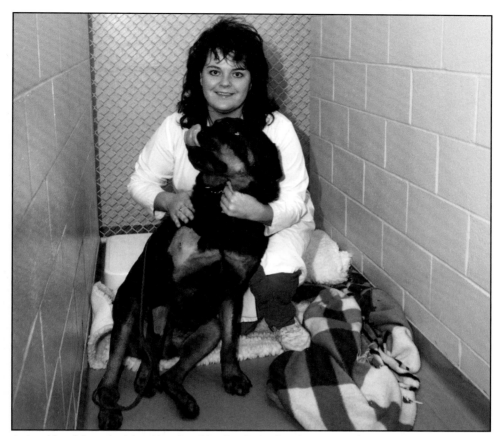

Animal health technician Heather MacSwain in the dog run with Bruno, a Rottweiler from Cape Breton, NS, who was recovering from knee surgery (c. 1994)

AVC animal scientist Dr. Mary McNiven (*right*) with J'Nan Brown and some of her goats

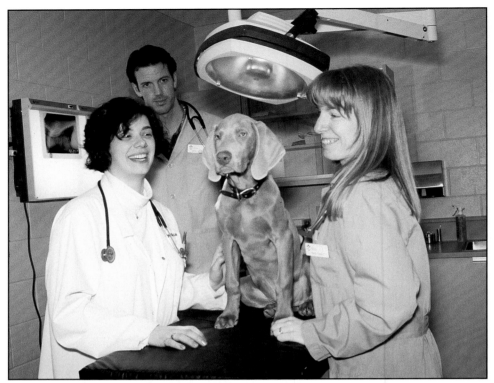

Dr. Kristi Graham with Ian Ross and Tracie Rundle-Lazar of the Class of 1999

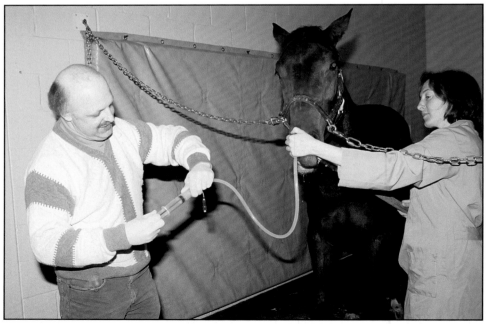

Dr. Ian Moore of the college's Equine Ambulatory Service being assisted by unidentified student.

Donna Barnes, animal health technician, with Sharp Shinned Hawk

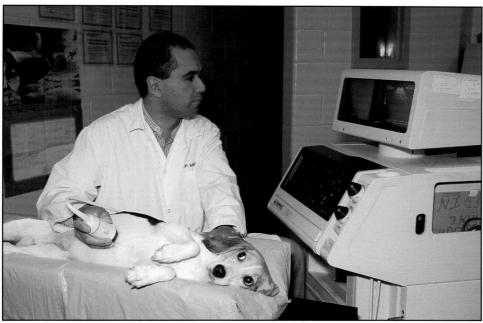

Dr. Mauricio Salano performing an ultrasound examination on a beagle

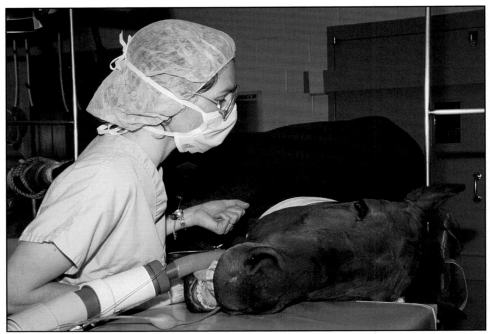

Technician Wendy Oliver with a horse in anaesthesia

Client Sonya Banks has her Nova Scotia Duck Toller admitted to the Veterinary
Teaching Hospital by student Beverly Greenlaw

premiers blamed the federal government.

At one point, a group of Charlottetown businessmen got into the act in an attempt to unstick the logjam. In May 1982, the Business Development Committee of the Greater Charlottetown Area Chamber of Commerce wrote to Whelan, proposing that a private developer finance all or part of the construction of the college, on condition that the Island government guarantee the funds.[3] This proposal was endorsed by the University of Prince Edward Island board of governors[4] and by Premier Lee, who wrote that this method of financing "may allow all the governments involved to reassess their positions and wholeheartedly support the location of the college in Charlottetown."[5] The flaw in that proposal, of course, was that the operating costs would be a much heavier financial burden than the construction costs.

Tom McMillan, a Conservative Member of Parliament whose Hillsborough riding encompassed Charlottetown, mentioned the possibility of private investment that spring in the House of Commons. At one point, he asked Whelan whether the school could go ahead without unanimous agreement by the four provinces. This exchange followed:

McMillan: "Will the minister inform the House whether the federal government is committed to funding a fourth veterinary college in Canada to be located in the Atlantic region? If so, does its commitment depend upon an agreement by all four Atlantic provinces on a site, or would an agreement by three provinces suffice?"

Whelan: "Madam Speaker, we made a decision several years ago. We even had $500,000 in the Blue Book to start the veterinary and aquamarine college to be established in the Atlantic provinces. They cannot agree among themselves. The terms of reference which I was given as the minister of agriculture were that if they agreed it would go ahead.... We are still waiting for them to agree. It is not our fault, that is for sure."[6]

McMillan brought up the issue again the next month. Did the federal government, he asked, consider it important to break the stalemate? Was the government pursuing the idea of private investment funding to get the college off the ground?

Whelan replied that an Island group had made such a proposal, had already met with government representatives, and was preparing a written submission to the government. He added, "But I think the tragic thing is that in the Atlantic provinces every province is under Tory administration and they cannot agree among themselves where it should be. I think that is the tragic part of it."[7]

Years later, McMillan confessed that he and Whelan, although on different sides of the political fence, had been in cahoots on this issue. They had conspired, he said, to argue about the college financing in the House, thus keeping the issue alive in the minds of the public and politicians.[8]

Through the years, talks on the veterinary college had taken place on two levels: among the politicians and among the non-elected officials. Barry MacMillan, a long-time public servant, was secretary to the Prince Edward Island Treasury Board in the 1980s. He recalled that Lee worked quietly behind the scenes, making many a phone call to his fellow premiers. And in the early 1980s, MacMillan and other staff from the Executive Council office made many a trip to Halifax, in fair weather and foul. "We'd hole up in a hotel for days. We were pushing the negotiations. We were more interested in an agreement, obviously, than they were. So it was not always easy to get people to the table. So we'd camp there for days at a time to see if there would be a break in the action."

The subject of the Agricultural College was always front and centre in those discussions. The Islanders kept trying to find ways of satisfying Nova Scotia by ensuring that a veterinary college in Charlottetown would not endanger the Truro college. "My sense was that there were some people [from the Nova Scotia Premier's Office] who simply wanted to get the issue behind them and move on to other things," MacMillan said. "We'd have discussions, and they'd go back and have discussions with the Agricultural College, and then we'd come back and talk some more. It was days upon days of this stuff. It was gruesome at the time."

A breakthrough finally came when the Maritime premiers met in Charlottetown in the spring of 1982. After the meeting, Buchanan was interviewed by CBC television host Roger Younker. Bob Nutbrown, Premier Lee's Principal Secretary, recalled that Younker asked Buchanan why he was refusing to let the veterinary college come to the Island, and that Buchanan replied, "I'm not stopping it. I just won't put any money towards building it."

Nutbrown said he then had a chat with Buchanan. "I said, 'Well, what if we build it?' He said, 'Yeah, I'd have to send money.' I said, 'You'd have to send money for capital and operations with each student, and you'd agree to a certain number of students.' He said, 'Yeah.' I said, 'Well, then, that's the deal, isn't it? ' He said, 'Yeah, could be.'"

Shortly after that, Nutbrown and Henry Phillips, secretary to the Island's cabinet committee on Intergovernmental Affairs, met with Buchanan in Halifax to get an agreement in principle to support the college project. Nutbrown speculated later that Buchanan had been primed by a prominent Quebec Conservative, Marcel Masse, who eventually joined the federal cabinet under Brian Mulroney. At the time, Masse was working for a consulting company out of Montreal. One day, while travelling through the Maritimes, he stopped in Charlottetown and had lunch with Nutbrown. The subject of the impending Prince Edward Island election came up. Nutbrown recalled, "He said, 'What do you think Jim Lee needs to win the next election?' I said, 'We need the vet college.' He said, 'I'm going to see John Buchanan tomorrow, and I'm going to tell him that.'" Whether or not that influenced Buchanan, he signed a letter of intent to support a veterinary college on the Island.

That may not have been particularly good news for the federal politicians. A few months earlier, Henry Phillips had discussed the college issue with Michael Kirby, deputy clerk of the Privy Council Office in Ottawa. Phillips reported to Premier Lee that Kirby had conceded that Nova Scotia's opposition was convenient for the federal government. "He admitted they [the feds] have an agreement to honour," Phillips wrote. "However, the government is in a time of restraint, and it would be easy for the Cabinet to turn down a request based on Nova Scotia's lack of support."[9] In June, Premier Lee went to Ottawa to plead the Island's case with Prime Minister Trudeau, and was told there were no federal funds available. And Whelan continued to blame the Nova Scotia government for the logjam.

"Prospects of Vet College for P.E.I. Don't Look Good," a headline in Summerside's *Journal-Pioneer* read on July 15, 1982. In the article, Whelan was quoted as saying that Nova Scotia's offer was not acceptable: Buchanan had agreed only to contribute to the tuition of students who wanted to attend. Whelan said Buchanan had to fall in line with the other Atlantic provinces. Furthermore, the newspaper reported, Whelan was starting to think the college would never be built: "I think a lot of people are sick and tired of talking about it, and so am I." Then the logjam broke — sort of.

On August 9, 1982, Lee and Buchanan signed an agreement to build the "Atlantic College of Veterinary Medicine and Animal and Marine Health Centre." The agreement read, in part: "Subject to the proviso that the Government of Canada shall commit itself to contribute 50% of the capital cost of the college, the Government of Nova Scotia agrees that the college shall be established in Charlotte-town. The Government of Nova Scotia undertakes that it will pay its appropriate share of operating, start-up and capital costs (amortized over a period of thirty years) of the college in accordance with the Council of Maritime Premiers formula...in return for a guaranteed annual allocation of a minimum of 16 seats per year...." Nova Scotia would pay 13.3 per cent of capital costs; New Brunswick, 10.85 per cent; Newfoundland, 8.925 per cent; and Prince Edward Island, 16.925 per cent.[10]

In other words, Buchanan was still not promising money up front. His political enemies therefore could not accuse him of selling out his province by sending money across the Strait to finance the college construction. Furthermore, he had lobbed the ball into Ottawa's court by hinging the agreement on Ottawa's commitment of half the capital cost. Buchanan had also removed some of the sting from his concession by upgrading the Agricultural College at Truro: the degree-granting program got under way in September 1981, and, in the early 1980s, the government was pouring money into the college to expand and build new facilities, including a library and an animal science building.

By this time, it didn't seem so urgent to locate the veterinary college in Truro, Buchanan recalled later, and people in the farm community who had lobbied hard for the Nova Scotia site were starting to "mellow." Besides, he said, his friendship with Jim Lee was a big factor in his decision, as was his recognition that Maritim-ers are all part of one big family. "We knew it would be helpful to PEI, but at the same time we knew that it would help the agricultural community in Nova Scotia, even though there were still pockets of discontent in Nova Scotia about it going to PEI. But, on balance, I think the consensus was, one, we should have a veterinary college in the Maritime provinces, and, two, whether it's in Truro or Charlottetown really doesn't make that big a difference. We're only separated by a bit of water."

The timing, of course, was right. Buchanan, politically astute as always, sensed that, despite the pro-Truro pockets, there would be no great public outcry among Nova Scotia voters if the veterinary college went to the Island. (He was proven right in the 1984 provincial election, when his party won forty-four of Nova Scotia's fifty-two seats, the biggest majority of any political party in fifty years.)

Unfortunately, the Lee-Buchanan agreement did not end the uncertainty over the fate of the college. "Veterinary College Again in Doubt," a *Journal-Pioneer* headline read on October 16, 1982. Whelan was quoted as saying that the level of Nova Scotia's support still did not satisfy the federal government. All the province had offered, he said, was to contribute on the basis of the number of Nova Scotians attending. In a reply that fall to a letter from Bennett Carr, Commission Chairman of the Village of West Royalty, Prince Edward Island, Whelan said, in part, "While a conditional agreement has been signed between Premier Lee of Prince Edward Island and Premier Buchanan of Nova Scotia with respect to the establishment of this institution, I have not as yet received from Premier Buchanan a statement of his unequivocal support. I consider this necessary before I make my presentation to Cabinet to seek approval for the commitment of funds."[11] In December, at the close of a Council of Maritime Premiers meeting, Buchanan challenged Whelan to "put his money where his mouth is."[12] A month later, Premier Lee was gathering correspondence among the provinces and other documents to try to convince Prime Minister Trudeau that Nova Scotia was, indeed, firmly committed to the project. "I don't seem to be able to get this across to the federal Minister of Agriculture," he lamented.[13]

Possibly it was a message that the federal Minister of Agriculture did not particularly want to hear at that time. The next year, Ottawa was still pinching pennies, and the money that had been set aside for the college construction had evaporated.

Still, Whelan recalled later, he tried to keep the issue alive in cabinet. His friend Dan MacDonald, the Island's strongest advocate

A few of the major players in the negotiations for a new veterinary college (*left to right*): PEI Premier Jim Lee, NB Premier Richard Hatfield, former PEI premier Angus MacLean, and federal Minister of Agriculture Eugene Whalen, at the official opening of the AVC, May 1987.

in cabinet, had died in 1980, but his influence remained. Whelan recalled a conversation with the Prime Minister: "So finally Trudeau said to me, 'Why do you keep bringing this up in cabinet?' I said, 'Mr. Prime Minister, we made a promise to Dan MacDonald that we'd build a school of veterinary medicine. The survey said it should be in Prince Edward Island, and we have a commitment from the provinces now.' 'Well,' he said, 'get on with building it.' I said, 'I don't have any money.' He looked at Lumley and he said, 'You have money.'"

Whelan was on good terms with Ed Lumley, the Minister of Regional Industrial Expansion, and it so happened that the two men were flying to Charlottetown together in February 1983. They made the deal to transfer the money from Lumley's department, Whelan recalled, "over some concentrated agricultural product" at an altitude of about 25,000 feet, on a Falcon aircraft travelling 400 miles an hour. "I waited until I had him where he couldn't escape, and where he was softened up for the knockout punch. I don't know whether I threatened to throw Lumley out of the plane or not."

Considering Trudeau's orders, Whelan's skyjacking was probably unnecessary, but it gave the two ministers a chance to share a good story about Lumley's largesse when the plane landed in Charlottetown: Ottawa was committing $500,000 to begin the planning and design of a new veterinary college.

That news took many Islanders by surprise. "It would be wonderful if we could read minds," an editorial in the Summerside *Journal-Pioneer* mused. "Then we would be able to find out what mental process occurred to change suddenly Agriculture Minister Whelan's attitude to the proposed veterinarian college to be located on the University of P.E.I. campus. A couple of weeks ago in Halifax he was maintaining that it wasn't his fault that the college was not proceeding, that it was all the fault of Premier Buchanan of Nova Scotia. Without advance notice at a meeting in Charlottetown on Friday, being held for another purpose entirely, he announced that the federal government had found a half-million dollars to be used to draw up plans and begin design work for the college."

Although the federal government had made that commitment, there was still a danger the deal would fall apart. The provincial governments still needed to work out the details of a formal deal to share both capital and operating costs. That spring, federal and provincial officials scheduled a meeting in Halifax. Lee gave his troops their marching orders: "I said, 'Whether it's two days, two weeks or two months, you're going to stay in Halifax until you have an agreement.' It did take a number of days, well over a week spent closeted in hotel rooms, and caucusing back and forth, until I got a phone call that an agreement had been reached that we would be satisfied with."

In May 1983, Whelan cemented the deal by announcing the federal government's commitment of $18.25 million, half the projected construction cost of the veterinary college. The dream was finally becoming a reality. The final agreement said nothing about Nova Scotia's spreading the capital cost over the next thirty years, but did contain a major concession to Nova Scotia on the Agricultural College.

Fortunately, at that time the premiers were in the mood to co-operate. One of the best symbols of this spirit of co-operation was the agency that had started the ball rolling on the college, the Maritime Provinces Higher Education Commission (MPHEC), established by the three premiers in 1973 and charged with representing each of the provinces and their collective university education interests. As Dr. Ken Ozmon, former president of St. Mary's University in Halifax, pointed out, the premiers did a number of deals in those days, sending the crime laboratory and MPHEC to New Brunswick, the office of the Council of Maritime Premiers to Nova Scotia, and, eventually, the veterinary college to Prince Edward Island. "There were trade-offs, and I think the vet college was one of them," Ozmon said.

The veterinary college agreement might be construed as an example of successful regional co-operation. But Alex Campbell, who had done his part for Maritime unity by helping create the Council of Maritime Premiers in 1971, said the long-drawn-out saga was also "an example of how difficult it is to make things happen in the Maritimes when you need the approval of all three governments. People in Ottawa must look down here sometimes and shake their heads."

Through all the political in-fighting, Premier Hatfield, always a strong proponent of regional co-operation, was undoubtedly shaking his head as well. Years later, he noted that the building of the college was a monument to what happens when governments work together. He added, "But it also should be a monument to remind us what happens when we don't — it takes much longer to achieve something that is really needed, and I hope we all learn from that."[14]

Of all the players in this political poker game, Eugene Whelan stayed at the table the longest. Except for a brief interregnum when Joe Clark's Conservatives took over in 1979, Whelan served as Agriculture Minister from 1973 until the sod was turned for the new college. The other political players changed frequently, no doubt to the discomfort of those who were charged with keeping the politicians briefed on the issue.

Alex Campbell had sown the seeds for the college project by campaigning for it in the 1970s, and finding the funds to pay Reg Thomson to sell and plan it. What made Campbell's dream come true in the 1980s was the conjunction of a number of fortuitous circumstances and personalities: Jim Lee's friendship with Buchanan, Hatfield, and Peckford. Buchanan's decision to risk alienating some Nova Scotia voters. Richard Hatfield's consistent, unqualified support. Trudeau's respect for Dan MacDonald. Whelan's friendship with MacDonald. And Jim Lee's bulldog-like determination to get the job done — and get it done now.

5: A College By Any Other Name…

On the night of June 2, 1983, Bob Nutbrown received some troubling news over the phone at his home in Charlottetown. Peter Connell, the federal deputy minister of agriculture, was calling from Ottawa, with a message that could hardly have come at a worse time. It was the night before federal agriculture minister Eugene Whelan and political representatives of the four Atlantic provinces were to meet in Charlottetown to sign an agreement that, finally, would launch the new veterinary college. As principal secretary to Premier Jim Lee, and later secretary to the cabinet, Nutbrown had played a significant role in negotiations leading to this climactic moment. It was the biggest deal he'd ever helped pull off, and he wanted to make sure that there were no unforeseen hitches. There were.

Nutbrown recalled this phone conversation:

Connell: "I'm phoning to say that Mr. Whelan is not coming to sign tomorrow unless you call it the Atlantic Veterinary and Aquamarine College."

Nutbrown: "What is 'aquamarine'?"

Connell: "I think it's a colour."

Nutbrown: "I think so, too."

Nutbrown knew that Whelan was keenly interested in fisheries as well as agriculture, but he wasn't sure how well the emphasis on marine life would sit with the other provinces, particularly Nova Scotia, which had its own aquaculture research projects. He phoned Joe Clarke, secretary to the Nova Scotia Executive Council. Sure enough, Clarke was not happy. Nova Scotia would not sign an agreement that called the college an "Aquamarine" Centre. Clarke also thought of aquamarine as a colour. Besides, he didn't want to see any territorial conflicts over fish. Nutbrown then phoned Bennett Campbell, a former Island premier who was federal Veterans Affairs minister. Late that night, Campbell phoned back to say he'd been unable to change Whelan's mind.

Bob Nutbrown, Premier Jim Lee's principal secretary

The next morning at about five o'clock, Nutbrown awoke to another horrible thought: because of his preoccupation with other details, he had forgotten to invite the New Brunswick delegation — quite a significant oversight. He quickly phoned Barry Toole, chair of the New Brunswick Policy Board, who said he wasn't sure whether he could get any members of the cabinet to Charlottetown on such short notice. He wasn't even sure he could get a plane, but he'd try.

Before leaving for the office, Nutbrown phoned one of Whelan's assistants in Ottawa. "I said, 'Can you tell me if Whelan's bluffing or not? I've got to know. I'm having a nervous breakdown here.' And he said, 'His plane is not on the tarmac, and it doesn't look like he's leaving.'"

The signing was scheduled for 11 A.M. at the Charlottetown Hotel. By about 9:30, the cabinet ministers had gathered in Nutbrown's office — agriculture ministers Prowse Chappell of Prince Edward Island, Malcolm MacLeod of New Brunswick, and Joe Goudie of Newfoundland; and Ron Russell of Nova Scotia's Management

Signing of original Five-Party Agreement in June 1983. *Left to right:* Malcolm MacLeod, NB Minister of Agriculture; Joseph Goudie, NF Minister of Rural Agriculture and Northern Development; Prowse Chappell, PEI Minister of Agriculture; Eugene Whelan, federal Minister of Agriculture; Jim Lee, Premier of PEI; Ronald Russell, NS Minister Responsible for Management Board; Bennett Campbell, former premier of PEI and federal Minister of Veteran Affairs.

Board.

Nutbrown made another call to Ottawa to suggest a compromise. The college could not have "aquamarine" in its name, but Whelan perhaps could speak at length during the ceremony about the aquaculture component of the college. The voice on the other end replied: "That's not good enough."

Whelan wasn't getting on his plane in Ottawa, and the Nova Scotia people were ready to leave Charlottetown on theirs. "We certainly weren't going to sign it with that name," Clarke said. "We didn't want to turn this thing into a real comic opera."

Then Nutbrown received another phone call, and another piece of bad news. Peter Meincke, president of the University of Prince Edward Island, told him that the Nova Scotia university presidents were sending a telegram to Premier Buchanan urging him not to sign the agreement. "I said, 'Peter, can I take this call in another room?' I decided not to tell anybody what was happening. First, I didn't believe bureaucrats could move quickly enough to get that telegram out. And also I had faith that Buchanan wouldn't go back on his word to Premier Lee.

"So now it's past the appointed hour, 11 o'clock. Premier Lee has the ministers come into his office, and he's chatting with them. I'm chatting with secretaries to cabinet. I'm sweating bullets. My chest is aching. It's heart attack time.

"So I said, 'Guys, I'm going to phone around and get you a box of lobsters.' So we got them all, ministers and deputy ministers, a big box of lobsters because Whelan still isn't coming and it's 11 o'clock. And I haven't told them he's not coming, just 'there's a little hitch.'"

Noon came. Still no Whelan. Then, just as the delegations were leaving for lunch, a secretary stopped them at the elevator: Whelan had just called. He would arrive in time for a four o'clock signing.

Whelan arrived, as promised, and after an hour or so of negotiations, the signing took place.

For Jim Lee, the signing spelled the major achievement of his term in office. For Nutbrown, it also signified a happy end to a stressful year and a couple of very stressful days. "We were euphoric," Nutbrown said. "We had negotiated our college for PEI."

The news, when it reached the media, also caused rejoicing in some Atlantic Canadian homes. For many young people, the news meant renewed hope for a career in veterinary medicine. Under the agreement, the college would provide spaces for forty-one students each year from the Atlantic region. Sylvia Craig of Halifax had always wanted to be a veterinarian, but had married young, and couldn't envision moving her family as far away as Guelph. By the time the agreement was signed in Charlottetown, she had a bachelor of science degree and two small children. "I can remember very distinctly being in the kitchen of my house, holding one of my babies, when I heard on the radio they were going to open the college," she recalled. "I remember thinking, hey, it's not too late! I could still go to veterinary school!"

Under the agreement, the new college would accommodate up to fifty undergraduate students in each year of a four-year program, and forty graduate students. Prince Edward Island was entitled to send ten students a year to the college; Nova Scotia, sixteen students; New Brunswick, thirteen students; and Newfoundland, two. In addition to those seats, nine seats would be marketed on a total

Sylvia Craig, of Halifax, NS, one of the first students to enrol at the AVC. She graduated in the Class of 1990, and was the first recipient of the R. G. Thomson Academic Achievement Medal for the highest cumulative grade average for the DVM program.

cost-recovery basis, preferably to students, or their sponsors, from developing countries. This was a provision very close to Gene Whelan's heart, and one of the selling points he made to Prime Minister Trudeau in seeking support to build the college. Whelan's idea was that the Canadian International Development Agency (CIDA) would pay for the costs of educating students from the developing world. The operating agreement stated: "Every effort will be made to provide veterinary training for candidates from developing countries and thus to help improve the food production capability in these countries."

The federal government would contribute $18.25 million for design and construction of the new facility, or half the total estimated cost. The four Atlantic provinces would contribute the other half. Prince Edward Island would pay 19.5 per cent and be responsible for all cost overruns; Nova Scotia would pay 16 per cent; New Bruns-

The Hons. Jim Lee and Eugene Whelan, two of the signatories of the Five-Party Agreement, on the occasion in September 2000 of the inaugural presentation at the AVC of the Honourable Eugene F. Whalen, PC, OC, Green Hat Award (named in honour of Mr. Whelan's favourite hat, which dates back to his days as a minister in Prime Minister Trudeau's cabinet)

wick, 13 per cent; and Newfoundland, 1.5 per cent.

A separate operating agreement, signed by the provinces, stipulated that Prince Edward Island would pay 38 per cent of operating costs; Nova Scotia, 32 per cent; New Brunswick, 28 per cent; and Newfoundland, 4 per cent. The agreement also contained several provisions designed to protect and enhance the Nova Scotia Agricultural College, including a letter from UPEI President Meincke to NSAC Principal Herb MacRae. The letter promised that UPEI would counsel students to enrol in a pre-veterinary program at the Agricultural College if their primary interest was farm animals; that the college would not give preference to UPEI students; and that, if the number of students in the pre-veterinary program at NSAC declined, the university would "take additional steps, in co-operation with NSAC, to attempt to preserve a consistent number of students entering the pre-veterinary program."

The agreement also said a twenty-member advisory council would be established, consisting of eight provincial government nominees, four provincial veterinary medical association representatives, one person each from the Nova Scotia Agricultural College and Agriculture Canada, four from provincial federations of agriculture, the president of UPEI or a designate, and the college dean.

Years later, Whelan said he had no recollection of why he held up the signing, or why he arrived when he did. Bennett Campbell speculated that, although Whelan had been determined to have the fish component of the college recognized in the name, he wasn't prepared to sabotage the agreement after all those years of negotiations. "So I think he vented his displeasure by keeping everybody waiting."

In fact, in years to come the college would become renowned for its work in aquaculture. And the "aquamarine" title stuck for a brief time. The documents the ministers signed simply called the proposed school a "veterinary college," which is the term Whelan used in his speaking notes for the ceremony. The next day, however, an article about the signing ceremony appeared in the Summerside *Journal-Pioneer.* The new college, according to the newspaper, was known as "the Atlantic Veterinary College and Aqua-Marine School." The Charlottetown *Evening Patriot* had another variation on the name: "the Atlantic Veterinary and Aquamarine College." Eugene Whelan would have been pleased.

6: From the Ground Up

Prince Edward Island Premier Jim Lee didn't waste any time. Once he was confident that an agreement on the veterinary college was imminent, he phoned UPEI President Peter Meincke. "Wherever Reg Thomson is," he told Meincke, "get him on the phone and get him back to Prince Edward Island, because we've got a project. We've got a veterinary college."

In Saskatoon, Reg and Helen Thomson were just leaving the house to go to dinner when the phone rang. It was Peter Meincke. Would Reg consider coming back? Thomson didn't give his answer immediately. Helen was not eager to uproot the family for the second time in less than four years. The Thomsons had sold their house in Charlottetown, had bought another one in Saskatoon, had developed new friendships, and were generally starting a new life. One of their daughters had enrolled in the University of Saskatchewan. Helen worried that the veterinary college deal might fall through once again. "But I also knew that if somebody else picked up the ball," she said, "Reg would be heartbroken because he had put so much time and energy into [planning and promoting the college]."

About three weeks after the five-party signing, Reg Thomson wrote his letter of resignation to the western college. He and Helen packed up their belongings, headed east, and bought another house in Charlottetown. In July 1983, Thomson began his new role as dean of the Atlantic Veterinary College.

Cliff Campbell, an engineer in the provincial Department of Public Works who had been working to get the college project off the ground, recalled that the Thomsons arrived back in Charlottetown on a Saturday evening. At 8 o'clock on Sunday morning, the phone rang in Campbell's home. It was Reg Thomson. "What are you doing?" he asked Campbell. "Can you come in to work?"

That call signalled the start of four years of intense labour on the largest, most complex building project that had ever been under-

Dr. R. G. "Reg" and Helen Thomson

taken in Prince Edward Island. The four-storey, 250,000-square-foot building would cost $36.5 million; include teaching, research, diagnostic, and hospital facilities; and eventually house more than five hundred students, faculty, and support staff, as well as a collection of sick and healthy animals. It was a project that had an inflexible deadline — about fifty students were scheduled to walk through the doors of the new college in September 1986 — and a tight budget. If the project went over budget, the Island government would have to make up the difference. Moreover, it was an undertaking that required considerable creativity on the part of the planners. It wasn't the first veterinary school in the world, but it was the first one in the region, and the first in North America to incorporate aquatic life to any great extent. And, during the construction years, the planners had to constantly remind themselves not to build in obsolescence: technology was changing so rapidly that the most modern equipment today might well be out of date by tomorrow.

Reg Thomson had already done his homework during his previous stint on the Island: he had written to veterinary colleges, government officials, scientists, and others throughout North America, seeking advice in planning the proposed institution. He had also

A number of the key participants in the planning, design, and construction phases of the building of AVC (c. 1985). *Front row, left to right:* Dr. R. G. Thomson, Dean, AVC; Dennis Clough, Director, Administration & Finance, UPEI; Dr. Ken Wells, former Veterinary General for Canada; Unidentified; Clive Stewart, Province of PEI; Gordon Weld, Chief Architect (Webber Harrington Weld); James C. "Jim" Johnson, Project Manager; *Back row:* Cliff Campbell, Project Engineer. Thomson, Clough, Wells, and Stewart made up the four-person project management team.

Key participants in the project management, design, and construction phases of the building of AVC (c. 1985). *Left to right:* Gordon Weld; Jim Johnson; Clive Stewart; Ken Wells; Judy Tassel, Administrative Support; Terry Davies, Construction Manager; Graeme MacDonald, Project Management Accounting

visited about fifteen veterinary colleges to collect information on curricula, staffing, floor plans, and equipment. By September of 1983, he had prepared a massive pre-accreditation report for the American and Canadian Veterinary Medical Associations, outlining in considerable detail the direction the yet-to-be-built college was to take in terms of staffing, admissions, curricula, and so on.

Now Thomson, Island government officials, and "programmers" — consultants hired to translate into physical space the various functions of the college — set out on a pilgrimage to investigate facilities at other veterinary colleges and at fisheries research centres. The team took photographs, toured buildings, and asked for advice from the top people in the field.

Reg Thomson was, as usual, totally dedicated to the task at hand. Cliff Campbell, who had by then been seconded to the project management team, recalled rolling into Saint-Hyacinthe one night at about 10 o'clock. Before the team members had a chance to retire, Thomson, as usual, called a meeting to plan the next day's activities and decide exactly what the group hoped to learn from the Saint-Hyacinthe College officials. When the meeting ended well after midnight, the weary team went to bed, only to be awakened by a great commotion in the middle of the night. Campbell, who was sharing a motel room with Glen Beaton from the Department of Public Works, was amazed to see Beaton's bed sliding toward the door. Campbell headed for the door as well. An earthquake was in progress. In the midst of the uproar, Beaton managed to see some humour: "Reg must be trying to start another meeting," he joked.

Reg Thomson's meetings were, in fact, most instructive. The tour opened the eyes of the construction team. "We'd never seen a vet college before," said Barry MacMillan, then Secretary of the Prince Edward Island Treasury Board, "although we'd been talking about it for years, and negotiating for it. What the heck did one look like?" MacMillan encountered issues quite outside his domain — the storage and use of animal cadavers, for instance. "Flooring was another big issue, which we'd never think of. When you're dealing with large animals, what kind of flooring works for horses over the long term so they don't slip and slide?"

Thomson was the only veterinarian on the team, so, in conjunction with these tours, he enlisted the aid of about two dozen outside experts, mostly faculty and staff from existing veterinary colleges, to help plan the new college. Assigning each of these "surrogate users" to a section of the college, he sent them plans the architect had already made, and asked for comments: What kind of space and equipment would they need if they were working at the college? A veterinarian at the Western College of Veterinary Medicine, for instance, was asked to take responsibility for the facility for the anatomy laboratory; a faculty member at the Ontario Veterinary College helped plan the ideal teaching hospital. Where possible, the architects incorporated the amended design into the plans, sent them back by courier, and waited for further feedback.

The planning phase was well under way by May 1984, when the sod-turning ceremony took place. As the flags of Canada and the four participating provinces snapped in the breeze, the audience shivered under the weak springtime sun, and a gaggle of politicians sat on the outdoor stage, including agriculture minister Eugene Whelan, wearing his trademark green Stetson, but not including a representative from the Nova Scotia government. Premier Lee explained that a rain-and-wind storm had prevented Ron Russell of Nova Scotia's Management Board from attending the ceremony. Don Downe, president of the Nova Scotia Federation of Agriculture, represented that province, wielding a gold-plated shovel and throwing his hardhat in the air, in concert with President Meincke and the politicians. Whelan gave a capsule history of the genesis of the college from Ottawa's point of view, concluding, "I have looked forward to this event for a long, long, long time." Bennett Campbell, Minister of Veterans Affairs and a former premier of Prince Edward Island, compared the history of the project to the weather that afternoon: "The [Howell] Report, when it was received, was received coolly. The discussions were brisk and breezy and long. But nevertheless, I guess one has to consider that even though this project was the most discussed, the most analyzed, the most reviewed, the most talked about — we nevertheless are very proud and pleased to witness the true beginning of the Atlantic Veterinary College here in Prince Edward Island."

At that point, someone who had just wandered onto the campus grounds might not have suspected that anything was happening, were it not for the red-tipped survey stakes on the soccer field next to the library. The college had to be ready to receive the first crop of students in a little more than two years' time. The most recently built veterinary college on the continent, Wisconsin's School of Veterinary Medicine, had taken six years to build.

The Prince Edward Island government was in charge of construction of the Atlantic college, and government projects were notoriously prone to take longer, and cost more, than predicted at the outset. Cliff Campbell recalled hearing from some skeptics among the four provinces that had a stake in the project. "There were a number of watchdogs from the other provinces who were kind of skeptical at the outset, about whether we could deliver the project on time, and an even bigger question, whether we could deliver under budget."

One factor that silenced some of the armchair critics was the presence of Jim Johnson at the construction helm. A highly respected engineer with experience in Ontario and PEI, he had just completed a large project for the provincial government on the Charlottetown waterfront — a hotel and convention centre, completed on time and under budget.

The other positive element was the project management team's decision to adopt a "fast-tracking" process. That meant that, instead of taking the time to plan the entire building in detail at the beginning, the project team would divide the work into forty-two phases, allowing planning and construction to take place simultaneously. Each phase had a deadline and a strict budget. The system meant that you might be planning the detailed design of a section of the building while the concrete was being poured for its perimeter. In describing the process, Jim Johnson used to employ the analogy of a small boy running down a hill at top speed: it was a dandy way to get to the bottom fast — so long as you didn't trip on the way down. "The big risk with fast-tracking," Cliff Campbell said, "is that you don't know what the final costs are until the last contract is awarded. The positive part of it was that we were able to break the project down into bite-sized contracts that would attract Island contractors. If we put out a $36-million contract, there wouldn't be a contractor on PEI that could bid on it."

The only phase that went over budget when the tenders came in was the structural steel contract. The planners had to redesign that phase to whittle down the cost, and then top up the budget with money that had been set aside in a contingency fund.

Tour of AVC during construction period (c. 1985). *Left to right*: Dr. R. G. Thomson; Unidentified; John TeRaa, PEI Energy Corporation; Jim Johnson, Project Manager; John Buchanan, premier of Nova Scotia

One design element that added to the cost was the underground section. About half of the building was to be built underground, mainly because the planners knew that an above-ground structure would dwarf everything else on campus. They wanted to keep the college in tune with the other buildings as much as possible. The basement would house clinics, diagnostic laboratories, holding areas for large and small animals, and some classrooms; the second level was to include the dean's office, classrooms, labs, and audiovisual and computer facilities; and levels three and four would house more labs and offices.

The actual construction began in August 1984. Construction workers, with backhoes, front-end loaders, tractors, and gravel trucks, moved onto the site that summer to build a parking lot that would hold building materials and machinery. Two months later, a crew started work on the southeast corner of the building, pouring concrete for a retaining wall, which was also the basement wall.

By then, the university had hired an assistant to the dean, Mel Gallant, a business administration graduate from Summerside who had worked in the private sector in Charlottetown and Montreal. His role was to serve as the college's chief financial and administrative officer. For the next two years, Thomson, Gallant, three of the four department chairs, and two secretaries shared office space with the construction group in two yellow bungalows located between the parking lot and the University library. The construction team included Johnson; Clive Stewart, a just-retired civil servant who was named chairman of the project management board; Cliff Campbell; accountant Graeme MacDonald; Dr. Ken Wells of the federal Department of Agriculture; and architects from the Webber Harrington Weld Group Inc., selected in February 1984 to design and construct the college.

"We had a lot of fun then," recalled Cora Conrad, then secretary to the dean. "It was kind of intimate. Everybody was so friendly, and we all got along together. It was really interesting to see the building coming together. It was like having your own baby." To lighten the pressure of work, the team threw parties from time to time. Bob Curtis recalled one occasion when he instigated a rare break in the middle of the day. He'd just received word that his second granddaughter was born, and, to celebrate, he brought a few bottles of rum, rye, and gin to the office, and invited the entire project team for drinks. Reg Thomson was in a meeting. When he arrived back at the bungalow, the offices were empty and a couple of phones were ringing off the hook. "We could hear Reg storming down the hall," Curtis said, "but by that time, I couldn't have cared less what he said. He came through the door: 'What the hell is going on here?' I had a glass all ready, about half-full of rum. I said, 'Drink that, it's good for you.' Most of the time he was very sober-faced, but he smiled, and as soon as he smiled I knew we were all right."

At the time, Thomson was carrying an unusually heavy burden on his shoulders. First, he had the full-time job of planning budgets, recruiting faculty, and working on curriculum, student admissions, and a host of other issues relating to the operation of the new college. An accreditation site visit team from the Council of Education of the American Veterinary Medical Association paid its first visit to the university in the summer of 1984 to determine what plans were in progress for such matters as finances, faculty, admissions, and facilities. (The college passed that first hurdle on the all-important

AVC construction, spring 1985

AVC construction, fall 1985

road to accreditation in November 1984, when the Council gave it "reasonable assurance" of accreditation.) In addition, Thomson was keeping a sharp eye on every aspect of the construction, arriving at Cliff Campbell's office every day with a pile of yellow "sticky" notes containing his list of requests, cautions, and questions. In October 1984, writing to Jean McDonald in the dean's office at the Ontario Veterinary College, he observed, "Things are going well here, and the bulldozers and scrapers are very busy. The prices are coming in on target or a little lower for the construction aspects." Others in the project team were under considerable pressure as well. At one point, the pace was so hectic, Cliff Campbell's family doctor ordered him off the job for a week because his blood pressure had shot skyward. Campbell stayed home for a couple of days, but then resumed his sixty-five-hour-a-week schedule. That was not Thomson's schedule, of course. "He was there in the evening when I left," Campbell said, "and he was there when I got back to the office the next morning."

One of the biggest challenges in the construction process was trying to keep one step ahead of changes in technology when it came time to buy the latest research and diagnostic equipment. In some instances, the technology changed so much during the years of construction that one piece of diagnostic equipment, for example, could do the work of three or four earlier models. "We had surrogate users that we could call up at any time, and say, 'What about this?'" Campbell said, "but we had to be constantly wary of building obsolescence into the project. Those particular users had a lot of experience, and they had obviously developed certain ways of doing things, teaching or research, but we had to constantly step back and say, 'Is that how we want to do it in five years' time?'"

Some of the rooms required more than two dozen services, such as medical gas, propane, hot water, cold water, de-ionized water, and filtered exhaust systems. In addition, there were huge waste-disposal problems to deal with — particularly chemical and radioactive waste. The hospital required all the specialized equipment of a human hospital, with the added snag being that it would have to deal with 2,000-pound patients prone to falling on hard-surface floors and thrashing about wildly when afraid.

At one point, the federal government threatened to complicate the process further by incorporating into the building a shelter for government officials in the event of a nuclear attack. The team in charge of building the college didn't think there was room in the budget for a fallout shelter, and both Premier Lee and Liberal Leader Joe Ghiz were cool to the idea. "We're not going to spend money on fallout shelters in this province," Lee told the Legislature in the spring of 1984. Ghiz said he was glad to hear that. After that, the idea sank without a trace.

One of Dean Thomson's pet projects was the front entrance of the building, which he felt should serve as an exhibition hall, rather than as office space. "When you walk into an expensive public building," he told a reporter, "you should see something." He oversaw the installation of display cases with exhibits such as skeletal forms, stuffed birds, and early veterinary instruments.

The biggest construction challenge arose in designing in detail the fish-health section, partly because the project team had received so much conflicting advice from surrogate users, who included people in the aquaculture industry and in the academic side of aquaculture. "I remember during the construction design phase," Barry MacMillan said, "nobody knew much about fish at that time. So the question was, how in heck are we going to build a water tank or a bunch of water tanks? Where are we going to get the salt water? There was some foolish talk about taking it from the Charlottetown harbour and dragging it out to UPEI. There was talk about these tanks being as big as a container truck." Eventually, Jim Johnson and Cliff Campbell sat down together one weekend, sifted through all the recommendations, and came up with a scaled-down tank system and provision for both fresh and salt water. A well was drilled outside the Fish Health area,[1] and a supplier was found to provide a mixture of salt and other nutrients that could transform fresh water into sea water.

Generally, the complicated process of involving surrogate users worked well. Dr. Bob Curtis, a large-animal specialist who joined the faculty while the college was under construction, considered that, on the whole, the college was well-designed from the point of view of faculty and students. For instance, the radiology department and the central supplies room were located conveniently close to surgeries for both small and large animals. But there was, Curtis said, the odd glitch. The surrogate user who gave advice on large-animal

Temporary quarters in basement of UPEI Physical Plant (c. 1985). *Back row:* Ian Dohoo, Arnost Cepica, Cathy Schaap, Kathy Mitton, Dawn Stillwell, Reg Thomson, Tim Ogilvie, Jim Bellamy, Floyd Trainor. *Middle row:* Evelyn Daley, Heather Cole, Sharon Derry, Shelley Ebbett, John Burka, Jamie Amend. *Front row:* Joan Darris, Cora Conrad, Carla Smith, Mel Gallant, Bob Curtis

surgery had recommended mobile tilting tables, and the college bought three of them, at $75,000 each. In theory, you could anaesthetize a horse, flip it onto a mobile table, and wheel it to the operating room. In practice, that didn't work very well. "Well," Curtis said, "we started interviewing surgeons to try to hire them, and you couldn't get a surgeon who would even look at those tables, wouldn't have anything to do with them. The tables weren't absolutely solid if a horse struggled and moved a bit. So we had to take that room and renovate it." That renovation involved installing a scissor lift in the floor onto which a surgery table is lowered. When a horse is anaesthetized, it is lowered onto the table, which is raised by the scissor lift, and the table is wheeled to the operating room. Curtis spent the next three years trying to peddle the old tables. He finally persuaded the Oregon supplier to take two back, and eventually sold the third.

Curtis planned the teaching hospital, with the help of Dr. Susan Dohoo for the small-animal clinic, and acted as adviser for the farm-service area of the college. At one point, he visited his former

Dr. R. A. "Bob" Curtis, founding chair of the department of Health Management, and Co-Director, Veterinary Teaching Hospital.

Dr. J. F. "Jamie" Amend, founding chair of the department of Anatomy and Physiology (now Biomedical Sciences), and Co-ordinator, Academic Affairs.

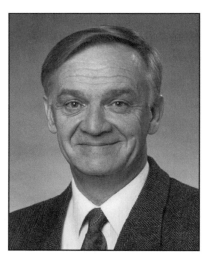

Dr. J. E. C. "Jim" Bellamy, founding chair of the department of Pathology and Microbiology, and Co-ordinator, Graduate Studies and Research.

Dr. Brian L. Hill, founding chair of the department of Companion Animals, and Co-Director, Veterinary Teaching Hospital.

workplace, the Ontario Veterinary College, to get some tips on a surface for the horse arena that could be easily cleaned, was reasonably dust-free, and would ensure good footing for the animals. The college at Guelph had installed used tires, chopped into tiny particles. It looked ideal to Curtis — until he learned that the people at Guelph had been ordered to remove the flooring because the tires contained PCBs. He scrapped the used-tire plan and substituted a mixture of sand and clay.

Curtis was one of the first three department chairs hired at the new college. He arrived in May 1985 to take over the department of Health Management. He and Reg Thomson had both studied veterinary medicine at the Ontario Veterinary College, and later had worked together at Guelph in a team-teaching situation. Each respected the other enormously. Curtis, a popular teacher with twenty-five years' teaching and farm-service experience, was a pioneer in preventive medicine, a field that Thomson was particularly keen on promoting in Charlottetown. Thomson gave Curtis free rein in setting up the farm-animal department, and also relied on him to help recruit faculty, a job with some challenges. "Reg was having real problems getting anybody to come here," Curtis said. "There had been some real opposition to the college, published in the CVMA [Canadian Veterinary Medical Association] *Journal*. That feeling was still there when we went to recruit faculty. The Americans weren't lining up, either." There was fierce competition for faculty at veterinary colleges throughout North America. And this was a brand-new institution with no reputation. Although the college had "reasonable assurance" in the fall of 1984 of future accreditation, there was no possibility of receiving full accreditation until the first class graduated in 1990. Some faculty outside of Canada, Curtis recalled, couldn't even find Charlottetown on the map. The biggest incentive the college could offer was the attraction of the Island lifestyle. "Cheaper housing, for example. And we'd take them out and show them the Island, tell them what a nice, quiet place it was — only one murder in the past ten years." For their part, candidates for faculty positions had to describe their experience, goals, and ambitions. They'd fly in on a Thursday, meet the university president, other faculty members, and the personnel officer on Friday, talk to real estate agents on Saturday, and give a lecture and undergo a three-to-

Among the first faculty members at AVC, Dr. Tim Ogilvie (on the telephone) and Dr. Gerry Johnson (at his desk) work in temporary quarters established in the basement of the UPEI Physical Plant during the construction of AVC, 1985–86

Technologists Geoffrey Paynter and Heather Briand, at work in temporary laboratory set up in basement of UPEI Physical Plant during construction of AVC, 1985–86

four-hour interview with the selection committee on Monday. "One problem we ran into," Curtis said, "is that sometimes by the time we reached our decision, they were hired by another college. It was a big job — draining, as far as I was concerned."

Nevertheless, by July 1, 1985, the college had appointed chairs of three departments: Curtis in Health Management; Dr. James (Jamie) Amend from the University of Nebraska in Anatomy and Physiology; and Dr. James "Jim" Bellamy from the University of Saskatchewan in Pathology and Microbiology. (Dr. Brian Hill was appointed head of the Department of Companion Animals in July 1986.) By the end of that year, seven more faculty had been recruited: Dr. John Burka, from the University of Alberta, in Anatomy and Physiology; Dr. Timothy Ogilvie, formerly on the faculty of the Ontario Veterinary College and more recently provincial veterinarian for the Prince Edward Island Department of Agriculture, in Health Management; epidemiologist Dr. Ian Dohoo, formerly with Agriculture Canada, in Health Management; Dr. Arnost Cepica, veterinary virologist with Health and Welfare Canada, in Pathology and Microbiology; Dr. Richard Cawthorn, from the Western College of Veterinary Medicine, in Pathology and Microbiology; Dr. William Stokoe, from the University of Edinburgh's Royal (Dick) School of Veterinary Studies, in Anatomy and Physiology; and Dr. Amreek Singh, from the Ontario Veterinary College, in Anatomy and Physiology. A dozen or more staff, including photographer Shelley Ebbett and graphic artist Floyd Trainor, and several secretaries and technicians, also joined the team that year. Every new faculty member received a warm welcome that included a dinner party prepared by Helen Thomson.

Bellamy, who had double duties as co-ordinator of Graduate Studies and Research and as chair of Pathology and Microbiology, recalled that those construction years were hectic for everybody. Besides helping plan and equip the building, he was setting up a graduate studies program, preparing his own courses and lectures, negotiating with government departments to set up contracts for diagnostic services, and recruiting faculty. "We were in meetings all day long, and probably worked until nine or ten every night for those first two or three years," he said. "But it was exciting."

As the hiring progressed, faculty and staff outgrew the yellow-bungalow offices, and the new arrivals moved to an L-shaped

Official opening of AVC, May 1987. *Left to right:* Richard Hatfield, Premier of NB; Dr. Gary Morgan, president, CVMA; Tom McMillan, MP for Hillsborough; Prowse Chappell, PEI Minister of Agriculture; Gordon Campbell, Chancellor, UPEI; Dr. Willie Eliot, President, UPEI. *Speaker:* Dr. R.G. Thomson, Dean

Unveiling of the plaque at the official opening of AVC, May 1987. *Left to right:* Dr. Hugh Whitney, Provincial Veterinarian, NF; Tom McMillan, MP for Hillsborough; Joe Ghiz, Premier of PEI; Richard Hatfield, Premier of NB; unidentified; Eugene Whelan, federal Minister of Agriculture

space, nicknamed "the dungeon," in the basement of the university's utility building. Faculty and staff sat cheek-by-jowl at desks in an open space with concrete walls and floors, girders on the ceiling, and the occasional whiff of exhaust fumes from the utility trucks rumbling overhead. For the next year and a half, they brightened up their quarters with posters on the walls, worked on research projects, prepared lectures and programs, and became acquainted with each other. "We were a very close-knit group," said photographer Shelley Ebbett, who was assigned to take pictures of everything from post-mortems to construction zones to students. "It didn't matter who you were working for — we were all in there together, and everybody, faculty and staff, partied together, went to the Christmas barn dance as a group. It was fun. Once we moved to the new building, that was lost."

While the construction was proceeding, some of the faculty were looking for research equipment. The new Queen Elizabeth Hospital had just been built in Charlottetown, and some equipment from the old hospital was being auctioned off. Dean Thomson, who was not only industrious but also thrifty, dispatched Bellamy to the auction to see what he could pick up. "We bought an autoclave and a few other odd things that helped the researchers get going," Bellamy said. "Reg was pretty pleased, because it didn't cost us much."

Meanwhile, the college was starting to interview prospective students. One complication in that process was that students had to be recruited for the nine seats set aside for people outside the Atlantic provinces. When Eugene Whelan, the long-time federal Minister of Agriculture, had begun promoting a fourth veterinary college in the 1970s, he had envisioned that the college would educate veterinary students from developing countries. In fact, he said later, it was on this basis that he had sold the college to Prime Minister Trudeau. Ottawa would finance the cost of educating these students by funnelling the money through CIDA. The agreement the four Atlantic provinces signed in June 1983 stated that nine seats in the college would be marked on a total cost-recovery basis every year, and that "every effort" would be made to reserve these seats for candidates from developing countries. The Island government didn't begin actively marketing those international seats until 1986, when it hired Charlottetown school teacher Mike Murphy to do the

Official opening of AVC, May 1987. *Left to right*: Gordon Campbell, UPEI Chancellor; Dr. Willie Eliot, President, UPEI. *Speaker*: Premier Joseph Ghiz, PEI

Student Jane Hogan registers for admission into the Class of 1990, the first class registering at AVC in September 1986, with Dr. Amreek Singh, faculty member; Carla Smith, administrative support; and Sharon Gormley, administrative support

job. By then, the federal government had changed hands — Eugene Whelan was no longer in the cabinet — and CIDA had changed its funding programs. It no longer funded individual students directly unless their education was part of a larger development program. For a few months, Murphy knocked on doors of overseas diplomats in Ottawa, but they weren't interested in paying the costs of a Canadian veterinary medical education. By the spring of 1986, it became obvious that the best hope of finding foreign students who could pay the $26,685 tuition fee — the cost for students outside Atlantic Canada — lay south of the border. Consequently, the college advertised in a couple of American newspapers, and the applications started rolling in.

Eleven American students were accepted in the first class. Among them was Dale Paley, a resident of Miami, Florida, who had worked as a veterinary technician for ten years, and had applied to veterinary school at the University of Florida. "I did not even get an interview," she said. Jane Hogan of Morell, Prince Edward Island, had also tried unsuccessfully to get into veterinary school. That was the year only one Islander was accepted at the Ontario Veterinary College. She was studying for a master's degree at Dalhousie University when she was accepted at the new college in Charlottetown. She was the first person in line when it came time to register.

On the morning of September 2, 1986, Hogan, Paley, and fifty other students — in total, twenty-six men and twenty-six women — assembled in the one classroom that was ready for use. As they arrived, Dean Reg Thomson was setting up chairs. Dr. Mary McNiven from Health Management, who was to give the first lecture (it was on milk marketing boards) smashed a bottle of champagne against the wall. With that christening, the first class of the first year was under way. Only one laboratory, histopathology, was ready, and its windows were covered with cardboard; the corridors were bare concrete; the Fish Health Unit consisted of only a bare concrete room; the teaching hospital would not be ready for another sixteen months; and students, faculty, and staff had to wear hardhats in some parts of the building. But Dean Thomson finally had the beginnings of a college — the first crop of students and eighteen faculty members, the core of a teaching staff that included some of the most highly respected teachers, researchers, and clinicians on the continent.

John Barrie (Fall 1986). John, a student in the Class of 1990, died in a whitewater rafting accident in 1991.

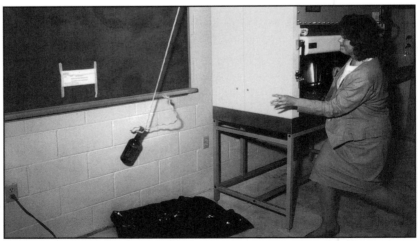

Dr. Mary McNiven, faculty member in the department of Health Management, smashing bottle of champagne to "launch" AVC, September 1986

However, by this time the dean had lost one of his greatest allies, Peter Meincke. As president of the university, Meincke had laboured hard from the beginning of his term to bring the veterinary college to Charlottetown. Meincke, like Thomson, was a scientist, and an ardent promoter of energy conservation and the appropriate use of technology. He saw the college as a vehicle for helping the university grow, enhancing its image among prospective students and forging ties with the Island's rural population. He and Thomson had liked each other from the start, and had grown close as colleagues and friends. But the recession had created some severe budgetary problems at the university; Meincke's administration had been party to some controversial budget-cutting measures, and he was not a popular figure in some parts of the campus. When his first term as president ended, he was offered only a one-year extension on his contract. Disappointed that he would not be able to oversee the fruits of his labours, he declined to accept that offer, and left his post in 1985.

Meincke's replacement was a man with a somewhat different outlook. C. W. J. "Willie" Eliot was a classics scholar who had taught in Greece, at the University of British Columbia, and at Mount Allison University in Sackville, New Brunswick. "I was as far away from veterinary medicine as you can get," Eliot said, "and that may tell you something. I definitely remember being asked during the interview process what I thought of the university going into veterinary medicine: 'Do you approve?' I said, 'That's not a question I ever have to answer because you've already approved it. If I come here, I've accepted that as part of the university. It won't be a reason why I would come here, and it won't be a reason why I wouldn't come here.'"

That response must have satisfied the selection committee. In July 1985, a little more than a year before the Atlantic Veterinary College opened, Eliot took over as the new president of the university, and oversaw the growing pains of the first few years in the life of the Atlantic Veterinary College. It would prove to be an exhilarating era — but a bittersweet one, too.

Dr. Brian Hill, faculty member and chair of the department of Companion Animals, with students of the Class of 1990 (Fall 1986)

Students Christine Mason and Jackie Rideout in the Class of 1990, with Ed Mason, invited lecturer on the lobster industry, from the provincial department of Fisheries and Aquaculture (Fall 1986)

7: Growing Pains

The new president could hardly have been more different from the new dean. Willie Eliot, son of a British Army officer, was born in India, lived briefly in England, moved to Canada with his Canadian-born parents in 1940, and was educated at private school and at Trinity College in Toronto. His academic interests were archaeology and classical Greece. Reg Thomson, a farm boy from rural Ontario, studied at the Ontario Veterinary College and Cornell University in Ithaca, New York. His passion was veterinary pathology. Eliot was an enthusiastic amateur pianist and singer with a sense of the absurd that was not always appreciated by his colleagues. Thomson had his own, quiet sense of humour, but was exceedingly conscientious about focusing on the job at hand. "Reg was second to no one in scholarship," said Lawson Drake, who was dean of Science at the time, "and, as an administrator, he was meticulous to the point of being fussy, and if you want to see his monument just go to the Vet College and look around. But he was a bit plodding at times, and Willie was prancing. And you get a plodder and a prancer side by side...."

The university president and the college dean may not have been soulmates, but there were other stresses and strains during the adjustment period of the marriage between the college and the university. For the first few months, construction was still going on. The college still had faculty to recruit, laboratories to prepare for the second year of the program for the Class of 1990, a teaching hospital to plan, a farm-client list to acquire, programs to assess and change. And always there was the issue of how the college fit into the larger university community.

Eliot said he respected Thomson greatly as a scholar and as the architect of a superb institution. "I would say that by and large we got along pretty well. There were difficulties. I would say the largest difficulty was that I was president and he was not. There was only

Unveiling of photograph of founding Dean R. G. Thomson, a work commissioned by the CVMA and donated to the college during the official opening of AVC, May 1987. Presented by Dr. Gary Morgan, President, CVMA

Dr. R. G. "Reg" Thomson, Dean, AVC, with Dr. C. W. J. "Willie" Eliot at the official opening of AVC, May 1987

Peter Scott-Savage and Sylvia Craig, students in the Class of 1990, with Dr. Ray Long, faculty member in the department of Pathology and Microbiology (Fall 1986)

one president, and it was not him. I think there must have been times when he found it difficult telling his staff they couldn't do things because the president wouldn't agree to it."

Reg Thomson was not a devious man, nor a manipulative one, and it is questionable whether he actually aspired for more autonomy — not for its own sake, at least — for his beloved veterinary faculty. But he was a well-focused man, and his focus then was on ensuring that the new school succeeded. Eliot, on the other hand, was determined not to let that school become the alpha dog in the pack. Eliot was exquisitely sensitive to protocol; he wanted to ensure that the college would not consider itself a semi-autonomous entity, but as a faculty subject to all the rules of the university and run by a dean who was subordinate to the president. He was also sensitive to the history of the university, with its roots in two small liberal arts institutions, and no doubt conscious of some of the fears that the intrusion of a professional school aroused among some faculty members. The college must serve to enhance the university, he felt, not overpower it. Even the name, "Atlantic Veterinary College," troubled him. It sounded to him as though the college had a life of its own. He always referred to it as "the faculty," never "the college." Trouble was, the college *was* unique. To pretend otherwise was like rounding up a herd of cattle and one elephant and insisting that the big guy with the trunk was just another cow.

The mere size of the building marked it as the elephant in the herd, even though it had been cleverly designed to slouch and sink much of its bulk underground so as not to tower over everything else on campus. The history of the college also made it unique: it was a prize that had been snatched, after much effort and with much fanfare, from a larger and more powerful sister province. The college had research and diagnostic facilities that the Science faculty could only dream of. The college was luring research-oriented faculty members to a university that, at that time, was primarily an undergraduate, teaching school. The college was developing — was required to develop — a graduate program. The college had its own crest, its own letterhead. Much to Eliot's dismay, the college was the only faculty on campus with its own cafeteria. Everything veterinary students needed on campus, except for books, could be found in their building, as self-contained as an ark sailing happily through the night.

Perhaps most significant of all were the college's separate financial arrangements. With a full enrolment of about two hundred students, the college had an operating budget of almost $15 million a year in 1990–91, compared with about $21 million for the university as a whole. About two-thirds of the operating funds came from the four Atlantic provinces, and the university president had little say over how the money was spent.

Then there was the matter of salaries. At the end of Eliot's second year on the job, he discovered that Dean Thomson was making more than he was. "One of the members of the board [of governors] asked me how my salary compared with the dean of Veterinary Medicine. I said, 'You'd better ask the dean of Veterinary Medicine what his salary is and I'll tell you what mine is.' There was a rather difficult silence, and someone said, 'The dean's salary is such and such.' That meant it was more than mine. His salary was all arranged before he came."

That was one reason, Eliot said, that Thomson felt that professors

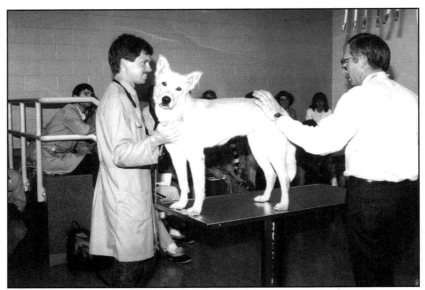

Tom Wright and other students in the Class of 1990, with Dr. Brian Hill, faculty member and chair, department of Companion Animals (Fall 1986)

at the veterinary college should also be paid a premium, if necessary, to come to Charlottetown. Thomson had little choice. Veterinary college faculty in some specialities were in chronically short supply. The only incentive a new, untried college could offer, in addition to a community with a bucolic lifestyle, was a salary competitive with other veterinary colleges across North America, or, in some cases, close to what a non-academic career could offer. Eliot understood this, and supported Thomson's request for a salary differential for veterinary faculty. Some of the members of the Faculty Association did not. Traditionally, the university administration determined salararies through a schedule, negotiated with the Faculty Association, of minimum salaries and increments. Those salaries, Lawson Drake said, were probably high enough to attract academics in other faculties, but not enough to compensate many of the specialists the veterinary college required. "Reg Thomson and others tried very patiently to explain this to the Faculty Association," he said. "The attitude of the executive at the time, and some others in the university, was, 'If you can't operate within the salary schedule, that's just too damn bad.'" To save the day, Eliot set up a committee to design a system whereby a candidate for certain faculty positions could bargain for a higher salary. In theory, this system could apply to some professors in other faculties, but, in practice, it meant that the veterinary professors were making more than most of their colleagues on the rest of the campus. "It was just a matter of supply and demand," said bacteriology professor Dr. Ray Long, who had taught at Guelph and worked as Provincial Veterinarian in Nova Scotia before joining the AVC team. "That was understood — but it wasn't liked."

A few people on campus, among them prominent members of the Faculty Association, had been suspicious of the veterinary college long before the sod was turned. Some felt that a professional school would compromise the university's identity as a liberal arts institution. A vocal minority fretted over the possibility that a veterinary college would skim off much-needed resources from the rest of the university.[1] By the time the college opened, most of the negative attitudes had changed. The Science faculty was particularly supportive. Science Dean Lawson Drake recognized that the college's extensive research program would inspire other faculties to press for better research facilities and a reduction in the teaching load, and that the Science faculty could participate in the college's graduate program, initiated in 1986, and eventually develop its own master's program.

Still, there remained a chill in some quarters toward the new kids on the block, and that puzzled and dismayed some of the college faculty. Drake recalled, "I remember Ray Long coming to me once in genuine puzzlement and saying, in effect, 'What's wrong with us?' All I could tell him was, 'You look good to me, Ray!'" Some people, like Drake, obligingly helped the new faculty members negotiate the treacherous rapids of campus politics. Some of the old-timers were only too happy to let the new boys sink or swim. Long remembered a contentious issue over which the veterinary college faculty held a point of view opposite to that of the Faculty Association executive. "At the Faculty Association meeting, the president of the Faculty Association could see how the vote would go. Suddenly he brought up a rule that we didn't have a quorum, so we couldn't hold the meeting, and we all went home. We found out later that there were plenty of people there for a quorum. But we didn't know that and they did."

In Lawson Drake's opinion, both sides may have suffered from over-sensitivity. The rest of the University faculty, who tended to

Theresa Bernardo, first recipient of a graduate degree (MSc) from AVC/UPEI

be teachers first and researchers second — if at all — may have detected, rightly or wrongly, the scent of academic snobbery from the research-oriented veterinary college professors. "The vet college people did come in with a lot of pride and expectation that some were a cut above the rest of us," Drake said. "This was a new place. They were especially hand-picked. They were good people and much in demand, and they knew it."

They also expected respect, and Willie Eliot did not always appear respectful. For one thing, he declined to address the veterinary faculty as "Doctor." He didn't care much for using "Doctor" for an academic degree, even the degree of Doctor of Veterinary Medicine. It was, according to the circumstance, either "Professor" or "Dean" or just plain "Mister." Some of Dean Thomson's colleagues felt that he, in particular, deserved to be treated with more deference by the university president. One incident loomed large years later in the minds of some of Thomson's friends. In May 1987, after the college was completed, a dedication ceremony was held in the large-animal exercise arena. Federal and provincial politicians, including the new premier of Prince Edward Island, Joe Ghiz, all had their say. Thom-

Ernest Hovingh, PhD (AVC's and UPEI's first PhD graduate)

son was last on the program. Eliot, acting as master of ceremonies, told the audience Thomson deserved as much praise as they wished to give him. But he also ordered the dean to cut short his remarks to save time, a directive which, Eliot observed, he had been unable to apply to the politicians. During his abbreviated remarks, Thomson confessed that the president's order had rattled him. It also rattled some in the audience, who felt that this should have been Thomson's day to shine. "That hurt people," Ray Long said. "That caused some hard feelings, because here the place isn't even officially opened and you've got somebody trying to hit you with a stick. That hurt, and it did a lot of harm and turned people off because Reg was so admired."

Some of the veterinary college students were too exhausted by the demands of their studies to pay much attention to the political climate on campus. Others, including Dr. Tom Wright, a member of the first class, felt a chill, rightly or wrongly, from the administration. "I remember I heard several speeches in which Eliot referred to the college, saying we were all in one big happy family, and that it wasn't us versus them. We were sort of rolling our eyes."

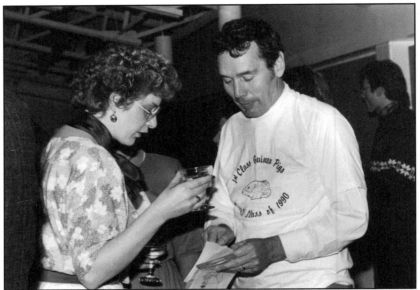

Ms. Rita Griffin, UPEI administrative support, with Dr. R.G. Thomson, Dean, AVC, sharing a moment during a social occasion (c. 1986)

That first class must have felt at times like the odd children of the big university family. The vet students were in their twenties, thirties, and forties. Most had one or two degrees. One had a PhD. They were the cream of the crop academically; they had to have an impressive record just to be considered for veterinary school. Some had been in the workforce for years, were married, had children. Like most students in graduate or professional school programs, they didn't have a lot in common with undergraduates barely out of high school. The veterinary students weren't at the stage where they were wondering what they'd do when they grew up. Some had waited for years for this opportunity. "Veterinary medicine is my passion," said Dr. Dale Paley, an American student in the Class of 1990. "If we have a purpose in life, this is mine: I was put on this earth to be a veterinarian." Paley, who was from Miami, Florida, had adjustment problems of her own. She hated Maritime winters (shortly after she arrived in Charlottetown, she spotted an ice-scraper in a friend's car and asked what it was used for; to her dismay, she soon found out), and she was frustrated by the slow pace of Island life, but she loved the school.

"They were the best and worst years of my life," she said. "It was tough. It was intense. There were many weekends when I went home and just cried. Because I was pretty intense about school, I was upset when they closed it because of bad weather in the wintertime. But I loved the vet school and loved what I was doing."

For Dr. Jane Hogan, those years were also "the best of times and the worst of times." She recalled wearing a hardhat in a makeshift anatomy lab in the cattle barn, as the construction crew worked around the students; trying to absorb an entire semester's worth of material from an itinerant instructor in four weeks; making her way through corridors lined with appliances still in their packing boxes. But she also remembered the intimacy of those early days. Students got to know each other and their professors, including Dean Thomson. He came to their house parties, and, after the first set of exams the first year, brought champagne to the as-yet-unfurnished cafeteria to help the class celebrate. "We saw a softer side of him," Hogan said.

Hogan's class was a special group, even in relation to those who followed. They called themselves "the guinea-pig class." They even had a T-shirt made with that slogan, and compiled a hand-lettered cookbook, titled *Guinea Pig Stew and Other Recipes,* to raise money for the class. "We were breaking totally new ground," Wright said. "We were learning as we were going, but so was the school. We sort of grew up together. We provided a lot of feedback. The faculty and the college looked to us to find out what was going right or wrong, and we were quite free and open to tell them." Because of the maturity of the class, Hogan said, the students weren't afraid to be critical. "We butted heads with a few of the professors. They didn't have nineteen- or twenty-year-olds sitting there. These were people with kids, people who had worked in other careers for perhaps ten years."

Because many of the veterinary students had been out of school for a while, Wright said, they had lost their competitive edge as students, and were eager to co-operate with each other. That helped nurture a spirit of camaraderie that was intrinsic to a group of people sharing the experiences of struggling to absorb mountains of information, worrying about exams, trying to find time to eat and sleep. Dr. Sylvia Craig, who led the Class of 1990 at graduation, recalled being somewhat disconcerted at the beginning of the second year.

"It was kind of hard when the new class came, because we'd almost forgotten that there were going to be other students there. We had a special relationship with the faculty, and we were a pretty close group. We certainly felt a sense of ownership about the college."

It was hardly surprising, then, that the Class of 1990 wanted to hold its own convocation ceremony. Eliot had been resisting efforts by the class to hold its own year-end ceremonies since the college opened. "I remember the very first year they were here, they had a great supper in the hotel, and they were going to give away various [university] prizes. I said, 'You're not. You haven't anyone to give them.' They said, 'The dean.' I said, 'No, the dean can't give away the university prizes unless either I or the registrar is present.' Then they said, 'Well, could you come?' I said, 'No. I have other things to do for the other people who are graduating.'"

Wright said the class wanted a separate convocation in a comfortable setting, such as the Confederation Centre of the Arts, so that elderly relatives would not have to sit on bleachers in the University gymnasium for four or five hours waiting for the veterinary students to go on stage. Besides, he said, there was a necessary restriction on the number of friends and relatives graduating students could invite to a university-wide convocation.

President Eliot stood his ground. "I said, 'Dammit, you're graduating from the University of Prince Edward Island. You're not graduating from the Atlantic Veterinary College. I'm going to give you your degree. No one else is going to. And if you don't come, you don't get it.'" During the convocation, however, the university president did acknowledge the distinct nature of the veterinary school grads. When it came time for them to receive their degrees, he took off his mortar board and donned a baseball cap sporting the AVC insignia.

That ceremony marked the conclusion of four years in which, despite the growing pains, faculty, students, and staff of the college had accomplished a great deal. The teaching hospital had opened. The Fish Health Unit was growing. The research grants were rolling in. The first graduate student, Theresa Bernardo, had received her master's degree. The college was gaining recognition in the outside world. The farm service had started. And, much to the relief of everyone concerned, in 1990 the college was granted full accreditation, that all-important stamp of approval from the Council on Education of the American Veterinary Medical Association, for a full seven years, the maximum period possible. This meant that the college met the essential requirements of a veterinary school as established by the Council.

And the campus climate was warming appreciably. One fortuitous event early on was the renovation to the Main Building just after the college opened. Many of the arts faculty had to move out, and spent a year working from offices in the veterinary college, where they made friends with people with whom they otherwise would have had little contact. As time went on, members of the Science faculty and the Veterinary faculty began co-operating on research and teaching. The older faculties began attracting more research funds and better research facilities. The Science faculty developed its own graduate program. Some professors retired and were replaced by people with no allegiance to any particular faction. The new, young faculty members tended to place more emphasis on research than did the old guard, so their priorities were more in line with those of the veterinary faculty.

And, despite the early adjustment pains, Eliot eventually concluded that the veterinary college had actually saved the university of Prince Edward Island. "Had the government been thinking of the possibility of getting rid of the university — after all, it was an expensive luxury, and there were plenty of universities in the Maritimes — now it would be very tough for the government to say, 'We don't want this university here.' The status of the university, the significance of the university, the worth of the university, everything was changed by the vet college. Here was a group of people from whom research, not teaching, was the essential. The university had a lot of people who did very good teaching but very little research. And I think that a lot of people suddenly saw, as a result of [the college], that people in anthropology, people in sociology, had a large place, had to get good faculty."

Eliot said the relationship between the veterinary college and the university was resolved to his satisfaction by the time he retired in 1995. The appointment of a new dean, Dr. Larry Heider, in October 1991, helped smooth the rough waters considerably. Heider, a graduate of the College of Veterinary Medicine at Ohio State University,

Dr. L. E. "Larry" Heider (dean from 1991 to 1998)

had a broad range of experience as a field-service clinician, as an extension veterinarian for dairy-cattle programs, and as an administrator. Before he joined the Atlantic Veterinary College, he had been chair of the department of Veterinary Preventive Medicine at Ohio State. In Eliot's opinion, Heider recognized the importance of symbolic gestures — using the university's letterhead, for instance, instead of AVC's. "He realized what I was getting at, that we all had to work together, as a single institution. And he did that well. He was always very, very careful about doing things with consultation and so on."

The unfortunate side of Heider's appointment was that it coincided with a much less happy circumstance — the end of academic life for the founding dean, Reg Thomson.

8: Farewell to a "Great Man"

During the official opening ceremony of the veterinary college in May 1987, Dean Reg Thomson publicly thanked his wife, Helen, for her support throughout the years of planning and building the school, noting that she "has put up with a severe case of the absent-minded, over-committed-professor syndrome." No doubt people in the audience were nodding their heads sympathetically. Since the late 1970s, Thomson had borne the main responsibility for planning, promoting, creating, staffing, and running a new, state-of-the-art institution that had to meet stringent, internationally set standards. He taught pathology to students at the college, and, in his spare time, he was working on another textbook. It was widely known that he had been working inhuman hours. He habitually left the college in the evening with not one, but two, briefcases in his bicycle carriers; evenings at home were for working on business correspondence. Nothing kept him from his responsibilities. His assistant, Mel Gallant, recalled trudging through the winter streets to the college during a particularly bad snowstorm, walking backwards in the biting wind. The whole town, including the university, was shut down. "I was thinking I was crazy to go to work," he said, "but when I got there, Reg was already in the office, smiling broadly when he saw me standing there, all caked in snow."

Reg Thomson had developed his work ethic early in life. As a boy he toiled long and hard on the family dairy farm in Ontario; if you were a Thomson, you were expected to work hard. His parents also expected him to bring home an outstanding report card from school. In high school, he always took on a full course load, never giving himself a spare period. And, even at that age, he exhibited a highly developed sense of right and wrong. That school had a fraternity that excluded Roman Catholics. Reg happened to be Protestant, but that exclusive club dismayed him so much, he founded another, all-inclusive club that did good deeds for older citizens in the town. In those days, he toyed with the idea of becoming a clergyman. In later years, he remained active in the church as a volunteer, but the career he finally chose was in academic life.

Thomson's students admired and respected him, although they did not always like him. He demanded much of himself, and of the people around him. Some students found him intimidating. "I don't think he set out to intimidate people," said Dr. Tim Ogilvie, a former student of Thomson who subsequently joined the AVC faculty, "but his knowledge of pathology was so complete, he intimidated people by the force of his knowledge. When you were in the classroom, you knew he meant business. He had a good reputation among people who were there to learn. Others didn't like him because he wasn't lighthearted at all." Thomson had an uncanny knack for spotting any ill-prepared wretch who couldn't identify a particular laboratory specimen, and he'd let the rest of the class know who that student was. That did not endear him, or pathology lab, to some students. But Dr. Tom Wright, a member of the AVC Class of 1990, said that as students became acquainted with Thomson, their respect for him grew. "He was bigger than life," Wright said. "He was a very imposing figure, tall, dark-haired, didn't smile much. He appeared almost cold at first. But when you got to know and understand him, you realized that he was really shy. And most of the time he appeared deep in thought. He was always trying to figure out to improve things, how to make the college better."

Thomson mingled frequently with faculty, staff, and students; he didn't hide in his office. He could be found in the corridors, the hospital, the labs, the barns, talking to staff, finding out what the problems were, how their jobs were going. Wherever he went, he habitually carried both a pen and a pencil in his right hand, apparently to save time. The pencil was for jotting notes on documents, the pen for drafting memos, signing letters, noting approvals. He had high expectations of his office staff, and he never failed to acknowledge extra effort. They were fiercely loyal to him. "Any extra work that we did, Reg noticed," said secretary Heather Cole. "There would be a rose on the desk, with a note saying, 'Thanks a bunch for a job well done.' We loved him. We thought he was such a great man. If ever there was a man who could put eight balls in the air and catch them all, it was Reg Thomson."

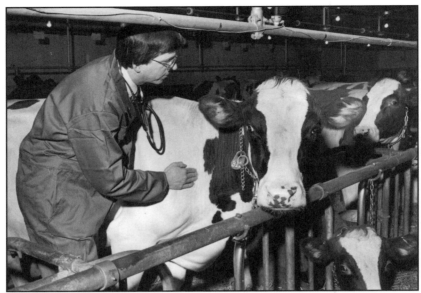

Dr. Tim Ogilvie with some AVC "teaching animals" in the college's north barn

In the late 1980s, Cora Conrad, the dean's secretary, noticed that he was becoming less adept at that juggling act. She ascribed it to the "absent-minded-professor syndrome" he had mentioned at the college's opening ceremony. "He would ask me for a file or something, and I would put it on his desk. A few minutes later, he would ask for it again. Of course, you never thought anything about it at the time, because you'd think, 'Oh, he's got a lot on his mind. He's a busy man.'" Mel Gallant came to a similar conclusion: "So many things were happening at once, and he was involved with so many files; if he forgot some little thing, you'd take that in stride."

Gallant and others at the college were relieved when Thomson decided to leave in May 1989 for a year-long travelling sabbatical, giving lectures in Britain, Australia, Malaysia, Indonesia, and New Zealand. His colleagues hoped that a year away from his many responsibilities at the college would give him a much-needed break. "He was so young and so bright and worked so hard," Gallant said, "I figured this sabbatical was just the ticket. I thought, 'He's going to come back renewed and invigorated, and he'll be fine.'"

Just before the Thomsons were to leave, Helen started having doubts about that. Suddenly, uncharacteristically, Reg stopped introducing her to people whom he knew and she didn't. "He seemed kind of fuzzy, not quite clear. I talked to him about it, but of course he denied that anything was wrong." Her premonition was right. The overseas trip did not go well. In Scotland, Thomson had trouble finding his way around the university campus. During a presentation in Ireland, just as he was preparing to show a set of slides, he picked up his papers and walked off the stage. On that trip, Helen noticed that Reg was having trouble with simple tasks. He would dial a phone number, abruptly hang up, try again, hang up again. He'd say there was something wrong with the phone. In fact, he couldn't remember the number long enough to dial correctly. Helen knew that Reg's father and uncle had suffered from Alzheimer Disease in their later years. But Reg was only fifty-six years old. "So I had this denial in my head: He was under a lot of stress. We were in a strange country. Everything was different. But I think deep in my heart I knew something else was going on."

The Thomsons returned to Charlottetown in May 1990, just in time to see the first class graduate. It soon became apparent that the sabbatical had not produced the results people had hoped for. At meetings, the dean took copious notes, and then had trouble reading them. He misplaced things. He mispronounced the names of students he knew almost as well as his own children.

One day, Cora Conrad went to Dr. Bob Curtis in tears. She knew something was terribly wrong with her boss. Curtis, Thomson's oldest and closest friend at the college, knew it, too. Thomson had appeared particularly confused at a meeting of Dean's Council, a body that consisted of the dean, his assistant, and the four department chairs. Curtis wanted his colleagues on the council to support him in confronting Thomson. They were reluctant to do so, partly, perhaps, because they weren't sure how bad his mental state was. Curtis knew his condition was serious, and that if he was no longer capable of doing his job, he should retire at once, with dignity. Curtis decided that, as a friend, he was obliged to speak to him. "Other than when my father was diagnosed with leukemia," he recalled later, "this was the toughest thing I ever had to do." Curtis went to Thomson's office and urged him to see a doctor. "Well, I was to a doctor last week and I'm in great shape," Thomson told him. "I

Dr. R. G. "Reg" Thomson and Dr. R. A. "Bob" Curtis at Laurie Blue's farm in Little Sands, PEI, in 1991. Mr. Blue hosted Draft Horse Field Days in which early farming techniques were demonstrated.

On the occasion of Dr. Thomson's receiving an honourary degree from UPEI in 1993 (*left to right*): Dr. Peter Meincke, former president, UPEI; Helen Thomson; Dr. R. G. Thomson, former dean, AVC; Jim Lee, former premier of PEI; Dr. Brian Hill, interim dean, AVC; Dr. C. W. J. Eliot, president, UPEI

don't mean that kind of a doctor," Curtis replied. "What kind do you mean?" "I mean a psychiatrist."

Thomson immediately phoned the doctor Curtis suggested — at that point, Curtis recalled, he was as tractable as a child — and got an appointment for the next Saturday. By Monday morning, Thomson, never a man to procrastinate, had decided to resign as dean. "That was a very sad time," recalled Dr. Ray Long, who had joined the veterinary faculty when the college opened. "He had done so much and had worked so hard. People at the college — faculty, staff, students — were literally in mourning."

Dr. Brian Hill took over as acting dean, but the college supplied Thomson with an office in the department of Pathology and Microbiology, and he continued to ride his bicycle to the college every day to sort through his papers. "He still had this dignity about him," Helen Thomson said. "He still went up there every day. I don't think I would have been able to do that. But the way he conducted himself amazed me, truly amazed me. He knew what he had. He knew what the future was. He was a pathologist. He'd seen his father and his uncle. But it wasn't 'Woe is me.' He just said at one point, 'Well, I got my mother's leg problems — his mother had thrombosis — and my dad's Alzheimer's,' and that was the end of it."

By the spring of 1991, the Thomsons decided to sell the house in Charlottetown and move back to Woodstock to be closer to their family. Before they left, a reception was held at the college to honour Reg Thomson. Interim dean Brian Hill, Premier Joe Ghiz, former premier Jim Lee, university president Willie Eliot, former president Peter Meincke, and department chair Bob Curtis all paid tribute to the former dean. "When the history of the University of Prince Edward Island is written," Eliot predicted, "there will be a chapter devoted to you." Curtis presented farewell gifts to the Thomsons, and described his former colleague as a man who always accepted a challenge, a man who was a good father and husband, and a man who could always be trusted. Thomson walked to the podium and said simply: "I don't have a big speech, but I just want to say to all of you, 'Thank you, thank you, thank you.'" Then he went back to his seat. Curtis and Thomson shook hands, and the two old friends embraced.

The Thomsons stayed at their cottage on the Island for another two summers. Then Helen decided to sell the cottage — "our last piece of PEI" — to simplify their lives. Reg returned for the last time in the spring of 1993 to receive an honorary degree from UPEI. During that visit, the college installed his portrait with those of the four original provincial veterinarians from the Atlantic region on a wall near the dean's office. The "Pioneers' Wall" — intended to honour the pioneers of veterinary medicine — had been one of Thomson's initiatives while he was dean. He unveiled his own portrait and was reminded that he, too, had been a veterinary pioneer in the Atlantic region.

The founding dean looked well, but, by then, his mental condition had deteriorated considerably. At one point, Dr. Larry Heider, who had been appointed dean in October 1991, invited him to his office to visit the staff. Heider had saved some personal effects Thomson had left behind. "We thought he might recognize them," Heider said, "but he had no recollection of them. I'm not even sure that he knew he was in his former office. But there was a very poignant thing that happened on that visit. Cora Conrad, who had been his secretary, was still there, and she was talking to him as we were walking out. He looked all around and said, almost from a distance, 'Cora, this is our old place, isn't it?'"

Thomson did not make a speech at the convocation or at a dinner and reception in his honour in downtown Charlottetown. At the evening dinner, Helen spoke in his place. She observed that her husband had never backed away from a challenge in his life; that he, in fact, thrived on challenges. But this was one challenge, she said, that he was not going to win. Despite that hard fact, Helen said later, the event turned out to be a joyous occasion for her and her husband. "It wasn't a downer. Dealing with Alzheimer's or any other disease, you go through the same emotions as you do when you're grieving over a death. There's denial, then anger. I used to be so angry. Then you go past that, to acceptance. So by the time we went back to PEI, probably I had reached that stage. Our life wasn't turning out the way we had expected, but then whose is?"

During those difficult times, Thomson received a number of other honours. One was an honorary degree from the University of Guelph, his old alma mater. Ray Long sought donations from all the retired veterinarians in the Atlantic region to establish the Dean

Reginald Thomson Fund to support AVC's graduate program. A scholarship was set up in his name for the top student in each graduating class. And, after he left the college, his influence remained in the programs he had developed and the people he had inspired. In the summer of 1990, shortly after he resigned as dean, Mel Gallant sent him a note. He praised Thomson for his more obvious accomplishments — helping break the political logjam to get the college built and overseeing its construction and growth phases — but he also spoke about less tangible matters. "Looking back," Gallant wrote, "I realize without a doubt that you have had a tremendous influence on any personal progress I may have made. I want to thank you for helping me to improve myself. I've often told friends that you 'set the pace' at the college. You work so hard, and you care so much, that you inspire people by your example. You influence them to be more than they have been. If there's a better definition of leadership, I don't know what it is."

The letter continued: "I learned a lot by working with you. I learned the wisdom of consultation and the strength of consensus. I also learned not to shy away from difficult decisions. You are the proverbial man of action. No issue is too small or too large to be tackled ... and there's no time like the present."

As Gallant observed, Thomson left the college in a position of strength. It had just been granted full accreditation for the maximum seven-year period, and the programs he had developed were starting to bring accolades to the college. Long before the college was built, Thomson had envisioned the growth of aquaculture in the region and the opportunity for a veterinary college to play a role in the industry. In the days when fish farming was in its infancy in the Maritimes, he used to travel to sea-cage sites in New Brunswick, and bring back photographs to show people such as Dr. Tim Ogilvie, who had worked as Provincial Veterinarian before joining the college faculty. "He was just like a kid with a new toy," Ogilvie recalled. "The light would go on in his eyes when he talked about the opportunities for veterinary medicine in aqua-sciences or fish health. Reg would say, 'This is a tremendous opportunity for veterinarians.' Being a traditional large-animal person, I didn't see the fit as well as he did." Thomson also recognized the importance of preventive medicine — probably before many practitioners in human medicine

Dr. Reg Thomson with Tiggs

did — and recruited to the college some of the leading lights in the field, including Bob Curtis, whom Ogilvie described as a "godfather" of preventive health care in Canada. Curtis, and the people he hired, subsequently turned on scores of students to the philosophy of prevention. "Even though Reg was a pathologist," Ogilvie said, "he was very good at understanding the breadth and depth of veterinary medicine. Some people focus on one discipline to the detriment of others. Reg was a mile wide and a mile deep."

Reg Thomson was a visionary with the fortitude to see his dreams come to pass. He was a teacher, a researcher, and an internationally respected author and editor. By the time he was in his late fifties, when he returned to Charlottetown to receive an honorary degree, he was unable to write his own name. By 1996, a decade after his college opened, his health had deteriorated to the point where he had to move to a nursing home. He died there on December 14, 2002.

His legacy — the institution he had worked so hard to build — went on from strength to strength. As the years went by, it helped forge ties between the university and the entire Atlantic region. And AVC was gaining acclaim at home and abroad in a way that even its founder may not have imagined.

AVC supplied this touch-tank during the 1999 Shellfish Festival

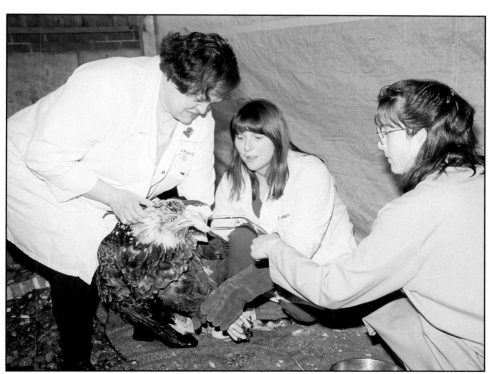

Dr. Caroline Runyon, animal health technician Donna Barnes, and student Barb Gilroy with an eagle

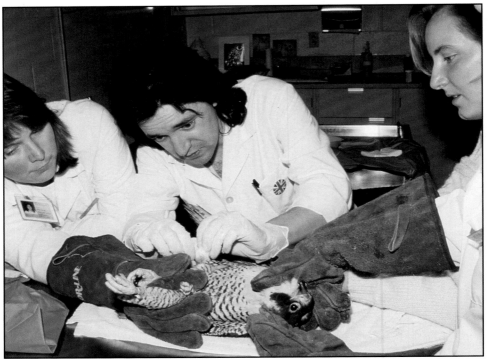

Dr. Tony Basher examines Peregrine Falcon with animal health technician Donna Barnes and student Jackie Rideout holding it in place

Dr. Art Ortenburger using acupuncture on a German Shepherd, with the assistance of animal health technician Heather MacSwain

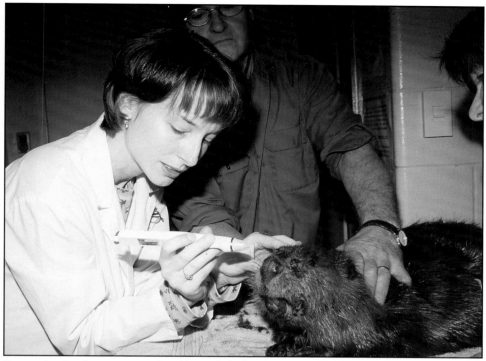

Beaver having its eyes examined by ophthalmologist Dr. Cheryl Cullen

A frisky equine patient romping about in the Teaching Hospital's paddocks

Dinnertime for these photogenic Ayrshires

Piglets

Dr. Trina Bailey with Dude and Ellie during the "Teaching Children About Pets Program" in 2000

In 1996, faculty and staff who had been at AVC since the college opened gathered to commemorate AVC's tenth anniversary

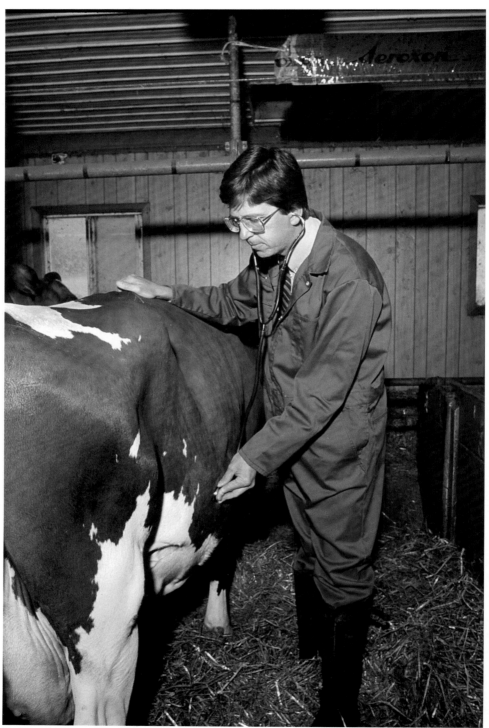

Dr. Tim Ogilvie with an Ayrshire

The Atlantic Veterinary College

Dr. Caroline Runyon releasing a Great Horned Owl

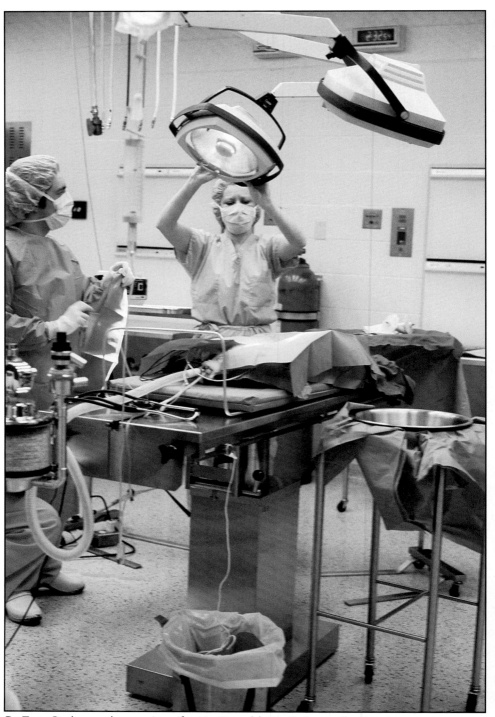

Dr. Tony Basher and nurse Jennifer MacDonald-Martin in small-animal surgery

9: Making Connections

One evening in mid-August of 1989, Marion MacAulay, manager of animal services at AVC's Veterinary Teaching Hospital, received an emergency call on her beeper: a dog named Duke, a German shepherd-husky mix, had been shot in the face in Yarmouth, Nova Scotia, and was being flown to Charlottetown. MacAulay met the plane, picked up Duke and his owners, Dale Duncanson and her daughter Meredith, and drove to the teaching hospital, where a surgeon treated Duke's wounds and eased his pain. His owners also needed some attention. They had left for the Halifax airport without even taking a change of clothes, much less reserving a hotel room. It was Old Home Week on the Island, the busiest week of the year for tourism. MacAulay spent an hour and a half on the phone before tracking down a bed-and-breakfast with a vacancy. The next morning, she drove the Duncansons to the AVC hospital to visit Duke, and then to the airport. A few days later, she loaded Duke on a plane bound for home.

For MacAulay, that kind of service was just part of a job that was much more than a job. In the early days of the veterinary college, she was the only staff employee in that section, and she did everything — bought cows, horses, sheep, goats, dogs, and cats, arranged for bedding and feed, cleaned stalls, fed the animals, and helped Dr. Tim Ogilvie, then a professor in the department of Health Management, teach a course in animal handling and restraint. On occasion, she operated a pickup and delivery service, at all hours of the day and night, for pets referred to the hospital from other provinces. In the early years, she was on call twenty-four hours a day, seven days a week.

For the Duncansons, the care and attention they and their dog received made the long trip by car and airplane worthwhile. Duke — "a magnificent dog" — had been blinded by the gunshot, and never recovered his sight. But there was nothing wrong with his constitu-

Ms. Marion MacAulay, manager of Animal Services in the Veterinary Teaching Hospital

tion, or his mental attitude. After he returned home, he steadfastly ignored his disability and lived happily for another twelve years.

At the time Duke became a patient, the hospital had been open less than two years, and the staff were working extra hard to make it a success. "We had such a sense of ownership about the place," said MacAulay, who later became Hospital Operations Manager. "There was a real sense of pride in being part of the process of getting the place off the ground. People never worried about their job description. We just did what had to be done." The efforts of those early years paid off. The hospital became one of AVC's most important conduits for gaining a reputation throughout the entire Atlantic community.

Over the years, the college managed to make connections in the Atlantic region and beyond through a wide range of activities under its mandate of teaching, research, and service. The Fish Health Unit,

Official opening of the Veterinary Teaching Hospital (January 11, 1988). *Left to right*: Dr. Brian Hill, chair, Companion Animals, and co-director, Veterinary Teaching Hospital; Dr. Michael Barr, president, Nova Scotia Veterinary Medical Association; Dr. Erwin Howatt, president, Prince Edward Island Veterinary Medical Association; Dr. Willie Eliot, president, UPEI; Dr. Bob Curtis, chair, Health Management, and co-director, Veterinary Teaching Hospital

the ambulatory service, the Animal Welfare Unit, the diagnostic and continuing education services, and a host of research projects all played a part. But the hospital was and is, in a sense, the storefront operation, the place most accessible — and most memorable, for good and for ill — to the general public.

There were some anxious moments in that regard. One involved a hunting dog from the Annapolis Valley in Nova Scotia. When he was admitted, he was so ill he couldn't even walk. "By the time he was ready to go home," MacAulay recalled, "not only was he mobile, he was feeling like a million bucks." Some of the AVC staff, heading for Nova Scotia early in the morning, and thinking to do a good deed, volunteered to take the dog as far as Truro, where they'd meet the owners. Unfortunately, while the staff were loading the van at 6 o'clock on a Saturday morning, the dog slipped away and ran off; his chauffeurs apparently had forgotten just how spry he'd become. Disaster! Somebody phoned MacAulay at home. She helped faculty and staff organize search parties to comb the city. The owners were notified. Flyers were posted all over town. Radio stations broadcast "lost dog" announcements all day. No dog. Finally, at about 10 A.M. the next day, someone captured the fugitive not far from the university, no worse for wear from his spree. After that experience, small-animal patients were walked only in a confined area outside the hospital.

That dog was one of many animals sent to the teaching hospital from off-Island. In fact, the first large-animal patient was Jock, a six-year-old Clydesdale horse from Hopewell, Nova Scotia. In laymen's language, he was a "roarer." His affliction, particularly common in heavy horses, occurs when a paralyzed laryngeal cartilage flops over the windpipe, partly shutting off the air supply, and creating a roaring sound as the air is drawn past the obstruction. His owner, Don Chisholm, made the four-hour trip by road and ferry to Charlottetown after a local veterinarian recommended the AVC hospital. At the college, equine surgeon Dr. Art Ortenburger repaired the faulty epiglottis, and, after about a week, Jock returned home. "He was a hundred per cent after that," Chisholm said. "The service was great. They treated him just like a person."

The referral caseload, from all four Atlantic provinces, grew from that point on. In 1991, 380 cases, including cats, dogs, horses, cattle, and pigs, were referred to the teaching hospital from other clinics, 154 of them off-Island. By the year 2001, a quarter of the approximately 5,300 small-animal cases in the hospital were referrals from other veterinary clinics. More than 940 referrals were from off-Island.[1] In addition, AVC clinicians and researchers shared their expertise over the telephone, free of charge, seventy to eighty times a week to practitioners throughout the region.[2]

Although the teaching hospital serves clients throughout the region, its primary purpose is to serve as an incubator for veterinary students — a place to integrate their academic life with practical experience. Students are introduced to the hospital early in their four-year program: in the first three years, they frequently visit the hospital, read records, and review cases, but can't do hands-on work with patients, except possibly in a minor way to assist a clinician or senior student while working part-time as a technician. Students in the early years also work with college-owned animals to get experi-

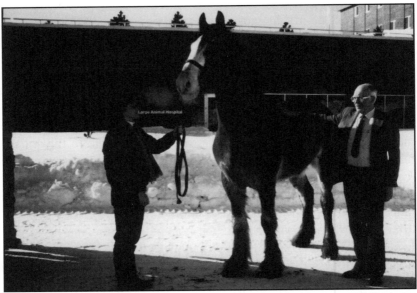

First case in Large Animal Clinic: Clydesdale gelding "Jock," with owner Bill Chisholm of Hopewell, Nova Scotia (holding reins), and Dr. R. A. Curtis, director, Veterinary Teaching Hospital (January 1988)

ence in handling, restraint, and examinations. In fourth year, students begin working at the hospital under the direct supervision of a faculty clinician. The rotations include examining patients, discussing diagnoses and treatment with the clinician in charge of a case, evaluating patients for anaesthetic procedures, working with various kinds of diagnostic equipment.

Like all veterinary clinics, and all human hospitals, the teaching hospital has had its share of successes and failures over the years. One of the drawbacks for clients has been that faculty clinicians have teaching and research duties in addition to their hospital shifts, which means that they're not always available at a moment's notice to discuss a particular case with a client. What the teaching hospital has to offer is a standard of care and a range of expertise unmatched elsewhere in the region, particularly for complicated cases. The medical staff includes specialists in ophthalmology, orthopaedic surgery, radiology, and anaesthesiology, as well as registered nurses, pharmacists, and animal health technicians.

And, from the beginning, the medical team has been equipped

with state-of-the-art diagnostic and treatment tools. At one time, in fact, the teaching hospital could do more accurate ultrasound diagnoses than could human hospitals on the Island. In the early years, cardiac specialists from the Isaac Walton Killam Hospital in Halifax used to travel to Charlottetown to care for Island paediatric patients at the college in the evenings. In the spring of 2001, the college acquired teleradiology equipment that would allow private practitioners to digitize images such as x-rays and send them to the college electronically. The hospital added an even more sophisticated procedure in January 2003: computerized radiographic equipment, in which x-ray images go directly on the computer for viewing, storage, and, if necessary, transmission to another veterinary school. In the diagnostic laboratories next to the hospital, technicians perform a wide array of tests, not only for clinicians at the teaching hospital but for practitioners throughout the Atlantic region. The large-animal arena contains an equine treadmill, which allows equine athletes to be exercised at racing speeds, so that clinicians can assess airflow obstructions, measure oxygen intake, analyze gait in slow motion, and correlate heart rate with speed. A gamma camera, an electronic instrument used to visualize the distribution of radioactive compounds in animal tissue, can detect the source of orthopaedic problems, such as lameness in a horse, when an x-ray cannot.

The surgical suites in the hospital were built, equipped, and maintained to a standard equal to operating rooms in a human hospital. Present during a typical operation, for example, would be a surgeon, an assistant surgeon, two or three students, an anaesthesiologist, one or two animal-health technicians, and a registered nurse. The nurse takes charge of procedure, ensuring that hospital policies are followed, that instruments are counted before and after surgery, that sterile techniques are adhered to. All the personnel are scrubbed, capped, gowned, masked, and booted. Sophisticated monitoring equipment enables the anaesthesiologist to keep track of blood pressure, heart rate, and other vital signs during an operation.

Dr. Wendell Grasse, who was appointed hospital director in 1997, said that standards such as these mean that a simple spay operation, for example, probably costs the college four times what it charges. One reason for setting the bar so high is to present an example for students. "It's in our best interest to teach the highest level of proce-

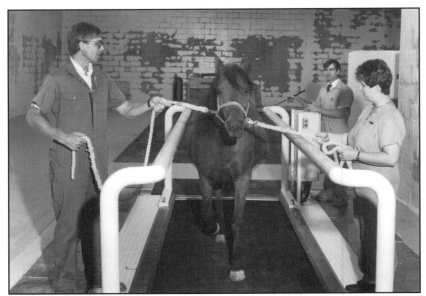

Treadmill in action. Left to right: Dr. John Pringle, Dr. Art Ortenbuger, and unidentified person.

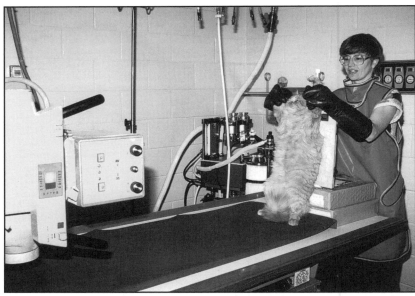

Dr. Patricia Hogan, radiologist and faculty member in the department of Companion Animals

Dr. Brian L. Hill

Interim Dean Brian Hill

Dr. Brian Hill was appointed acting dean of the college in May 1989, when Dean Reg Thomson was on sabbatical for a year, and served as interim dean in 1990–91, after Thomson resigned.

Originally from California, Hill graduated from veterinary school at Colorado State University, and taught at Iowa State University before moving to Prince Edward Island. He joined AVC in May 1986 as founding chair of the department of Companion Animals and eventually served as co-director of the teaching hospital.

Following his two years as dean, Hill returned to his position as chair of the department of Companion Animals, and later enrolled in a residency in radiology. Like many other veterinary colleges, AVC had been experiencing difficulties in recruiting qualified veterinary radiologists. Unfortunately, Hill became ill soon after beginning his program. He died on January 15, 2000, at the age of fifty-one.

During his tenure as dean, Hill made a number of significant contributions to the school. One was the establishment of the AVC Equipment Replacement Fund. Using "salary slip-

page" dollars from years when faculty recruiting was slower than anticipated, he put money aside for a day when expensive medical and research equipment would need replacement. By the year 2002, this interest-bearing fund surpassed $1.8 million. It was also through his initiative that the university agreed that the college should benefit from interest earned on college operating funds, a practice worth about $200,000 a year to the school.

Under Hill's leadership, the college adopted an innovative final-year program consisting of ten three-week elective rotations. Dean Tim Ogilvie said the concept of totally elective rotations was unique at the time among North American veterinary colleges.

In the opinion of Assistant Dean Mel Gallant, Hill's greatest contribution may have come simply through his personality. "He was funny in a self-deprecating way, and comfortable with almost everyone," Gallant said. "His calm, friendly demeanour, and his pragmatic, common-sense outlook on issues, greatly contributed to the college's effective integration into the larger university community. His early death was a great loss to the college."

Farm Service Group (1995). *Left to right*: Ricky Milton, technician; Dr. R. A. Curtis, chair of the department of Health Management; Dr. Ian Dohoo, faculty member in the department of Health Management; Dr. Dan Hurnik, faculty member in the department of Health Management; Dr. Jeff Davidson, faculty member in the department of Health Management; Dr. Mike Slana, faculty member in the department of Health Management; Dr. Ernest Hovingh, PhD candidate; Dr. Larry Hammell, faculty member in the department of Health Management; Dr. Patty Scharko, faculty member in the department of Health Management; Theresa Andrews, technician.

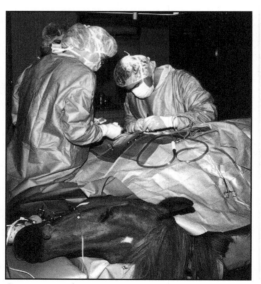

Surgery on a horse

Dr. Wendell Grasse, director, Veterinary Teaching Hospital

Verna Lee Dalziel admits clients with Marven MacLean looking on, as Allan Keoughan works in the background

Teaching horses grazing in the "off-season" at Winsloe Farm (AVC's off-campus facility)

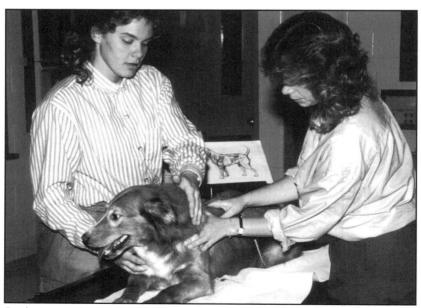

Students examine a patient in the Veterinary Teaching Hospital

dures, sterility, and so forth," Grasse said, "because students leaving here are in general going to drift downwards for the most part. When they leave here, the natural tendency is to gravitate to what is practical." In addition, the small-animal surgical suites must be ready to receive human patients in the event of a disaster that would overtax local human hospitals.

Although the teaching hospital cares for its share of sick and injured animals, the focus from the beginning has been on preventive health care — a position not much in evidence at one time in human medicine. Dr. Reg Thomson, the founding dean, believed that preventive medicine would be the cornerstone of the education of veterinarians in the future. "Salvage of sick animals has little or no positive impact on health and production in the animal industry," he wrote in 1979. "Medical treatment of sick individuals has little, if any, impact on the overall health of the community — human or animal."[3] Thomson's colleague Dr. Bob Curtis, who was director of the teaching hospital in the early years, shared a similar philosophy. Curtis thought that the image the hospital presented to the public should reflect that view. When the hospital was being designed, he insisted that drugs not be displayed in the admitting room. He wanted to make the point that a veterinarian's principal income should come from professional services, not drugs. "I knew we were going to have to sell some drugs, but the client would have to go to the counter to get them."

The emphasis on prevention extends to on-farm care of cattle, horses, sheep, goats, swine, and fish. Shortly after the college opened, it bought a number of farm animals to teach students restraint, handling, and basic examination procedures.[4] To recognize symptoms of a sick cow, for example, students first have to learn what a healthy one looks like. But the college also needed a caseload of large animals to teach herd-health management — such aspects as nutrition, housing, reproduction, and disease prevention — as well as diagnostic and therapeutic procedures. Consequently, in 1987, the Prince Edward Island government bought the Charlottetown Veterinary Clinic, a mixed-practice clinic with farm clients throughout the central section of the province. That purchase gave the college access to dairy, beef, swine, and sheep operations.[5]

The newly acquired caseload allowed faculty members to take

Curious porcine specimen

small groups of third- and fourth-year students to farms in the district formerly served by the Charlottetown Clinic to demonstrate procedures such as pre-breeding examinations, pregnancy diagnoses, post-calving examinations, nutritional evaluations, calf vaccinations, castrations, de-horning, and mastitis prevention. Most of the visits were uneventful. But one memorable day a farmer asked Curtis for help with a cow with milk fever. The cow was lying on a cement floor, and Curtis knew from experience that the only way to get her on her feet was to move her onto the grass where the footing was better. To move the cow, the students tipped her onto a piece of plywood, which the farmer attached to his tractor by a chain. Curtis told two of the students to hold the cow's head down to keep her on the makeshift stretcher. "I said to the guy, 'Just back up and we'll get her on the grass.' Well, I don't know what got into him, but he'd just bought a new tractor, and instead of putting it into ordinary reverse, he put the darn tractor in fast reverse. Well, the cow went out the barn door and over a pile of rocks at the end of the cement, and the cow went flying off, the students went flying off, and I was hollering at the farmer all the time to, for God's sake, slow down!" Despite the

Dr. Dan Hurnik (Health Management) with some busy piglets

fact that a female student was caught briefly under the cow, the story had a happy ending. The student wasn't seriously hurt, and, after a few minutes on the grass, the cow picked herself up, apparently none the worse for wear.

Wrestling with cows, performing rectal palpations, sloshing through the mud in coveralls and rubber boots — that was one way of translating the theory of preventive medicine into gritty practice. Clicking a computer mouse was another. Shortly after the college opened, it pioneered a service that, if lacking the excitement of flying cows, gave farmers a unique tool in herd-health management. In the mid-1980s, while the college was still under construction, Dean Reg Thomson wrote to the federal Department of Agriculture about an idea for a computerized information network that would help decrease the cost of livestock production. It would collect and merge animal-production records, slaughterhouse data, and disease records, and allow producers to compare the productivity and health of their animals with those from similar farms. It would also let producers pinpoint the timing of negative or positive changes in health or productivity, and evaluate the effect of new feeds or drugs. Epidemiologist Dr. Ian Dohoo, who had been working for the federal government, joined the AVC faculty, and in 1986 designed a computerized system to receive, analyze, and communicate information on herd health to farmers, veterinarians, and others. The system, eventually known as the Animal Productivity and Health Information Network (APHIN), started as a research project with dairy and hog producers on Prince Edward Island, and evolved into a fee-based service for hog farmers on the Island and, for five years, in Ontario. In the late 1990s, APHIN began disseminating information on the internet. The original system was cloned several times. After the salmon industry in New Brunswick was hit with a virus in 1996, project manager Barry Stahlbaum and other staff at APHIN designed a similar surveillance system for fish. The Quebec hog industry developed an information network that, in Stahlbaum's opinion, was modelled on APHIN. And by 2002, Stahlbaum was helping the lobster industry design its own network. For some farmers, APHIN proved useful in ways that may not have been envisioned by its inventors. At least one farmer discovered that he could literally take his herd-production graphs to the bank. "One hog producer wanted to expand his operation, build a new barn, increase the number of sows," Stahlbaum recalled. "To convince his banker that he was a growing concern, he took his APHIN graphs to the bank, and I heard that he got his loan, partly because of the graphs." Another farmer, contemplating the purchase of a neighbouring hog farm, asked to see the farmer's APHIN records. "After looking at them, he decided not to buy. So knowledge is power. That was the original goal — to put information into the hands of managers so they could make better-informed decisions."

At the same time as AVC was making connections to pet and livestock owners, it was also forming ties to a burgeoning aquaculture industry in the Atlantic region. The idea for a marine section at the college dates back to the 1970s. In 1974, the Prince Edward Island Department of Fisheries had begun investigating a system of growing mussels on lines suspended in the water of bays and estuaries. Four years later, the first crop of cultured mussels was harvested on the Island's eastern coast,[6] and New Brunswick started its salmon aquaculture industry a year after that with one farm and $40,000 worth of fish.[7] By the late 1970s, Premier Alex

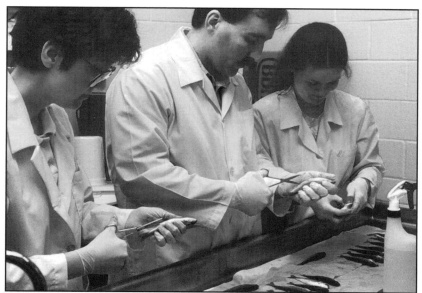
Technologists Nancy Hitt, Neil McNair, and Joanne Daley, in the lab

Dr. Gerry Johnson, faculty member in the department of Pathology and Microbiology and director, Fish Health Unit, with Paul Lyon, manager, Fish Health Unit, demonstrating fish-handling techniques to students in Class of 1990 (Fall 1986)

Dean Larry Heider

Dr. Lawrence E. "Larry" Heider served as dean of AVC from 1991 to 1998, a period of significant growth at the college.

Born and raised on a farm in Ohio, Heider graduated from Ohio State University in 1964, spent two years in private veterinary practice, and returned to the university to teach, his primary interest being dairy health management. While at Ohio State, he was appointed extension veterinarian, responsible for dairy cattle, and later became director of the teaching hospital and chair of the department of preventive health medicine.

Assistant Dean Mel Gallant, who worked with Heider during his tenure as dean, described him as a natural-born leader. Appropriately enough, upon his resignation from the university, a veterinary student leadership award was established in his honour. "He was a highly focused and energetic dean," Gallant said, "with a capacity for instilling confidence and a sense of purpose in his co-workers. He lived and breathed the true leader's confidence in the successful resolution of evolving challenges. Although the substantive message in the news conveyed at the annual 'state-of-the-college' meetings with faculty and staff was sometimes bleak, Larry's sense of direction and his strong hand on the tiller were enough to provide employees with a renewed sense of energy and purpose, enabling most of them to walk out of these meetings feeling better than they felt when they arrived."

Gallant recalled that Heider was also profoundly fond of his work and of his co-workers. No matter how busy he was, or how stressed the atmosphere, his usual response to the arrival of a colleague at his office door was to greet him or her with a smile and a handshake.

Heider's forceful personality may have intimidated some, but his openness to change and to dissenting views brought many new ideas out into the open. When some of his initiatives failed to fly, he accepted these setbacks philosophically — observing that there is wisdom in faculty decisions — and moved on to other projects. And he knew how to get things done.

As dean, Heider led the development of several new projects, including AVC Inc., the Animal Welfare Unit, the Canadian Aquaculture Institute, and the acquisition of the Cardigan Fish Hatchery. Heider also oversaw the establishment of the college's PhD program in 1996, and reorganized AVC's administrative structure. In the early years of the college, the faculty and founding dean had established a system of co-ordinators for various programs. Heider decided that one way to give greater prominence to these roles on the national and international stages was to establish the positions of associate deans for academic affairs and for graduate studies and research.

During Heider's term as dean, the premiers of the four Atlantic provinces reaffirmed their support for the college by signing the 1996–2001 interprovincial funding agreement. That fall, the college received the news that it had received full accreditation for seven years — the highest rating — from the Council of Education of the Canadian and American Veterinary Medical Associations.

In August 1998, after UPEI president Dr. Elizabeth Epperly resigned for health reasons, Heider was appointed acting president for one year. With his second term as dean scheduled to end in June 2000, he elected not to return to that position after completing his appointment.

In the spring of 2002, Heider left Prince Edward Island to take on a new challenge as executive director of the Association of American Veterinary Medical Colleges, based in Washington, DC.

Dr. Larry Heider, fishing at the Cardigan Fish Hatchery

Campbell was already thinking in terms of a marine section in the proposed veterinary college. In a letter to Prime Minister Pierre Elliott Trudeau in September 1977, he observed: "We are particularly hopeful about developing a strong program in aquatic medicine in association with other institutions in the region and the possibility of a small animal clinical facility in one of our major urban centres."[8] Irwin Judson, who had worked on the Island's oyster and shellfish program, was appointed Director of Aquaculture for the province in 1978. Shortly after he began his new job, he read in the newspaper that Reg Thomson had been hired to promote and plan a fourth veterinary college in Atlantic Canada. "We had biologists who knew something about fish," Judson said, "but what aquaculture needed was some veterinary science. That was quite *avant-garde* thinking at the time. Anyway, I called Dr. Thomson and told him that, in PEI, we all grew up with horse-and-cow doctors, and there were also pet doctors, but what we needed were some fish doctors. He said, 'I had exactly the same idea.'"

In his first month as planning co-ordinator, Thomson, outlining his vision for the proposed new school, suggested that veterinarians

Diagnostic Services manager Dennis Olexson demonstrating equipment techniques and capabilities to PEI Premier Pat Binns (centre) and his executive assistant, Peter McQuaid

Diagnostic Services

When Dennis Olexson was recruited from the Western College of Veterinary Medicine to set up the diagnostic laboratories at AVC, he insisted on calling the unit "Diagnostic Services" — with the emphasis on service. That meant, first of all, that the fifteen diagnostic laboratories would be integrated into one unit, with one central receiving station, to maximize efficiency. It also meant strict quality control and fast turnaround. "Anybody who knows me knows that I'm extremely fussy," he said. "If I'm going to do something, I'm going to do it right or not at all. Also, I'm a pet-owner — I love my pets — and just because it's an animal [that's being diagnosed] doesn't mean we can't do quality work."

Once the diagnostic unit was set up, it took over the work of Prince Edward Island's provincial veterinary laboratory, began providing virology services for livestock owners in the four Atlantic provinces, and eventually developed a client base of private veterinary clinics throughout the region. By 2003, the laboratories were performing about 360,000 tests a year for about 500 clients.

When those clinics send samples to AVC for analysis, they can usually count on a report in twenty-four hours or less. That fast turnaround time is made possible by the use of efficient analytic equipment, a fast delivery service, and a computerized reporting system. The college has a contract with a courier service that guarantees overnight delivery from the four Atlantic provinces. In an emergency, some test results can be processed within fifteen minutes to half an hour. Normally, though, when samples arrive at the lab before 9 A.M., the report goes to the client before 12 noon. Biopsies, which require a longer processing time, take an extra forty-eight hours. Reports are faxed automatically to the clinics, which also have the option of viewing their own reports by computer.

In 1997, the unit began serving clients far beyond the Atlantic region. Working in partnership with UPEI's Computer Services Department and Diagnostic Chemicals Ltd., Olexson initiated a quality-assurance program for other veterinary laboratories and hospitals. The program has attracted clients in Canada, the United States, South Africa, several countries of Europe, Australia, and New Zealand. Clients include most North American veterinary colleges. Each quarter, AVC sends quality-control samples to subscribing laboratories and hospitals, providing assessments for eight laboratory disciplines — bacteriology, chemistry/toxicology, endocrinology, hematology, histopathology, parasitology, serology, and therapeutic drug monitoring. Subscribers then can compare the results of their tests with those of their peers. "The program allows laboratories to monitor the quality of their work and track inefficiencies in their methods or instruments using results from their peers as a guide," Olexson said.

could make a significant contribution to the prevention, diagnosis and treatment of diseases in aquatic species. Island Fisheries staff had said that disease investigations and control programs would be needed for salt- and fresh-water trout rearing, as well as lobster and shellfish culture. "Considerable discussion will occur in each of the provinces on this general subject," he wrote. "It is one of the most futuristic concepts related to a new school and offers great potential for contribution for such an institution."[9]

In Thomson's view, an inhabitant of the sea should be regarded, for veterinary purposes, as just another animal. When the college opened in 1986, that viewpoint permeated its teaching, research, and service functions. Pathologist Dr. Gerry Johnson, a specialist in veterinary aquatic studies from Western Canada, was the college's first Fish Health co-ordinator. His job was to introduce veterinary medicine to the small but growing industry of aquaculture: in the Atlantic region, the industry by then was worth about $7 million in Atlantic salmon, rainbow trout, Eastern brook trout, blue mussels, and oysters.[10] But few veterinary colleges in North America were paying much attention to aquatic species. "Veterinary medicine and fish had been on divergent pathways up to this point in time," Johnson said. "We took on something outside of veterinary medicine and maintained it inside veterinary medicine. It was a big, big step."

On the teaching side, that big step meant that AVC students would study fish-related material in anatomy, physiology, immunology, and bacteriology, as well as safe handling and examining practices. In their fourth year, students would accompany a clinician on fish-farm visits, just as they did on trips to dairy and swine operations. "It's the teaching aspect that's unique about this place," Johnson said. "When you were learning the anatomy of the lung, you learned the anatomy of the gill, and when you were learning the micro-anatomy of the liver, it didn't matter if it was a fish or a chicken or a cow or horse or dog."

When Johnson arrived with the first students in September 1986, the Fish Health Unit, one of the last sections of the college to be completed, was little more than an empty concrete space. The aquatic facilities included a "wet" laboratory designed to provide a controlled environment for fresh-water and marine finfish and shellfish. Two fresh-water wells at the college supplied water, which would be

Dr. Gerry Johnson (*right*), faculty member in the department of Pathology and Microbiology and director of the Fish Health Unit, with unidentified mussel farmer

used to replicate sea water for salt-water species. Water temperature and flow would be regulated by computer-controlled mixing valves and monitored electronically. None of this equipment, however, was in place when the first class arrived. For student laboratory work, Johnson improvised by taking oxygenated containers with live fish to the anatomy lab.[11] By the time the college was officially opened in May 1987, modular tanks were installed in the aquatic holding facility, but they weren't quite ready to be stocked with fish. Johnson wanted guests at the opening ceremony to see live fish when they toured the new building. In a last-minute flurry of activity, Johnson and Fish Health Unit manager Paul Lyon activated the wells and plumbed the tanks for drainage and incoming water. They then arranged for rainbow trout to be delivered half an hour before the tour began. "I could just see the trout keeling over because we hadn't tested the water to that point in time," Johnson said. "But they were a big hit. People walked in, and there were fish swimming around. It was about the only place where there was a lot of live activity."

Those trout were still alive three years later, and, by that time, the Fish Health Unit had picked up speed like a fast train heading out of the station. By the spring of 1987, the unit had begun doing diagnostic tests on marine species. There was no shortage of clientele, to say the least. In fact, the response so overwhelmed the two people in charge of the unit at the time, Johnson decided the service had to be limited to farmed fish. "We were getting everything in here, sharks, wild fish, cultured fish. People would find sharks in their nets or washed up on shore and would want to have them diagnosed. We ran a diagnostic service. So the instant response was, 'We'll take them to that new Veterinary College.' And we just couldn't handle it."

The diagnostic service at first was limited to bacteriology and histopathology. As the unit acquired more staff, the service encompassed a complete set of diagnostic capabilities — post-mortems, bacteriology, histopathology, virology, endocrinology, haematology, mycology, parasitology, toxicology — everything that was available to diagnose illnesses in land animals. By 1989, two years after the service began, the laboratories were diagnosing more than eight hundred cases a year.[12]

One event that earned gold stars for the college in the early years was the domoic-acid crisis that shut down the shellfish industry in Atlantic Canada in the fall of 1987. In November and December of that year, more than two hundred cases of food poisoning, linked to contaminated mussels, were reported, mostly from Quebec consumers. Johnson joined a government-led task force to determine the nature of the toxin and its source. Once domoic acid was identified as the toxin, Dr. Louis Hanic of the UPEI Biology department pointed to the phytoplankton Nitzschia as its source. Working with Hanic, Johnson and his colleagues collected specimens and tested hundreds of samples that winter for the presence of domoic acid in algae from mussel-producing areas. "Louis was hauling in the algae samples and testing them on mice," Johnson recalled. "I can remember being in here on New Year's Eve — I had young kids at the time, and it ruined that whole holiday for me — and Louis had such a sore back, he was lying on the lab floor to rest his back, and I was running the samples into these mice to see what we had." Johnson finally was able to give Island mussel growers the good news that their beds were back to normal. And, eventually, AVC researchers collaborated

A two-coloured lobster (Homarus americanus), a genetic oddity

Lobster Lore

Until the mid-1990s, scientists in the fish health section of AVC worked mainly with farm fish. Then the lobster industry asked for help from the college in dealing with post-harvest losses. With the aid of a grant from the Max Bell Foundation, the college set up the Lobster Health Research Centre in 1996. Four years later, the college received a $3.3 million grant from the Atlantic Innovation Fund for lobster health research at the Centre, now called the Atlantic Veterinary College Lobster Science Centre (AVCLSC). Funding partners also included fisher organizations, processors, exporters, First Nations, and various governments.

The funding partners oversee research projects, which are designed to contribute to the health of individual lobsters and the economic yield and sustainability of the industry. "When we began working on lobsters," AVCLSC director Dr. Rick Cawthorn said, "my concern was, could we work on lobsters at a veterinary school, could I entice my colleagues to work on them, and could we answer questions from the industry." Since then, he said, about a third of the college faculty have contributed in some way to lobster research, and the Centre has been able to assist the industry. "The challenge," he said, "is to apply 150 years of traditional veterinary medicine, based on cats and dogs and horses, to an industry that is huge — the largest fisheries industry in Atlantic Canada."

Projects have included an examination of host-pathogen interactions in lobsters and the potential impact on the Canadian industry of infectious diseases such as paramoebiasis, a suspect in the die-off of lobsters in the Long Island Sound. The Centre also constructed a lobster research database for the industry. Technological innovations developed at the Centre include an ultrasound probe that determines meat yield in live lobsters and information labels for lobsters and other products. Cawthorn said research eventually will be extended to crab and shrimp.

with governments and the shellfish industry to conduct ongoing surveillance of mussel-producing areas.

By the end of the 1990s, the aquaculture industry in Atlantic Canada had reached the $225-million mark, four times its value a decade earlier, and more than thirty times its value when the college first opened its doors. New Brunswick's farmed-salmon industry was worth about $110 million, and Prince Edward Island's mussel industry about $20 million.[13] As the industry matured, its members recognized that the road to prosperity was paved with prevention. AVC, in line with its philosophy of preventive health care, had shifted its emphasis in the marine section from diagnostic to clinical services. Clinicians were making routine and emergency visits to farms raising fish, just as Health Management faculty did for cows, horses, hogs, and sheep. The aquaculture industry was looking for clinical and research partnerships. And the college was hatching a small but growing number of fish health specialists, ready to fill that niche.

One of them was Dr. Larry Hammell, who had graduated from the Ontario Veterinary College and, as one of AVC's first graduate students, had obtained a master's degree in fish epidemiology. Hammell joined the AVC faculty and eventually took over as co-ordinator of the Fish Health Unit from Johnson, who was appointed director of Diagnostic Services. Besides his teaching and research duties, Hammell provided on-farm service and health-management advice for fish farms, most of which were salmon farms in New Brunswick. In the early years of the 21st century, Hammell was on the road seventy to eighty days a year, taking fourth-year students to New Brunswick salmon hatcheries and sea-cage sites. From the beginning, he tried to get fish farmers thinking in terms of preventive health care, but, because aquaculture was so new, even the specialists didn't always know the risk factors for fish mortalities. Hammell began concentrating more and more on ongoing applied research at the fish farms, and on surveillance aimed at halting the spread of such diseases as the deadly infectious salmon anemia virus, which caused losses in the tens of millions of dollars a year to the New Brunswick industry. By 2003, AVC researchers were still attempting to develop a vaccine to combat the disease. Researchers were also studying a myriad of issues relating to the lucrative shellfish and crustacean industries. Dr. Jeff Davidson, for example, headed the Shellfish Research Group, which examined such problems as oyster die-offs on the shores of Prince Edward Island, and the invasion in coastal waters of the clubbed tunicate, a species that threatened the mussel industry.

To meet ever-expanding challenges in the industry, AVC's marine section had undergone a structural change in the mid-1990s. A variety of activities — teaching, research, continuing education, diagnostic services, clinical services — had functioned under one umbrella. In 1995, the unit spawned five separate units, each of which continued to grow in size and complexity. They included the Canadian Aquaculture Institute, which provides continuing education and training in aquaculture medicine, fish health, and management; the Aquatic Animal Facility, a laboratory designed to provide controlled environments for freshwater and marine finfish and shellfish; Aquatic Diagnostic Services; and Atlantic Fish Health, which does contract research for suppliers of products to the aquaculture industry. In addition, a number of research groups had sprung up to serve various sectors related to marine industries — the Lobster Science Centre; a Centre for Marine and Aquatic Resources, a research centre that collaborated with the UPEI Biology Department; and a for-profit corporate arm of the University, AVC Inc.

The genesis for AVC Inc. dates back to the late 1980s, when the college began offering contract research services to pharmaceutical and vaccine companies. In 1994, Atlantic Fish Health Inc., a for-profit company, was set up to do fish research, primarily contract research for the pharmaceutical industry. The for-profit company was designed for efficiency: instead of having to weave their way through a cumbersome university hierarchy to negotiate a contract, private firms could deal much more quickly with a single corporation. Soon, other opportunities arose for research. For example, the Department of Fisheries and Oceans wanted to divest itself of the Cardigan Fish Hatchery, which produced salmonids for fish-enhancement programs, and provided research, education, and consulting services for the Atlantic aquaculture industry. Because of the potential for research projects, the college didn't want to see the hatchery shut down. It was brought under the umbrella of the university company, now known as AVC Inc., two divisions of which were the Canadian Aquaculture Institute and a non-marine section, the Pork Production Innovation Group, which was operating a 1,000-swine facility and conducting research in nutrition, manure management, behaviour, and housing.

It had been hoped that profits from AVC Inc. would buoy the college's operating budget. Unfortunately, in spite of considerable benefits to the college from AVC Inc.'s expertise, collaboration, and worldwide activities and exposure, not all of the for-profit company's activities proved profitable, the company had to downsize somewhat by 2004 to stem the flow of red ink.

By the mid-1990s, countries around the world were encouraging the growth of aquaculture. More and more of AVC's continuing education, diagnostic, and consultative work was taking place internationally. The Canadian Aquaculture Institute was giving courses in Canada, the United States, southeast Asia, and Australia. In the summer and fall of 2002, AVC scientists travelled to Chile, Brazil, New Zealand, and some European countries. Jeff Davidson attended a conference in China. Gerry Johnson spent so much time in the air, "I don't have to carry identification when I go to the airport." AVC's reputation as "the fish school" was growing around the world — and not just with farmed fish.

As the years went by, the college gradually took on other marine species. Conservation officers and the general public from the four Atlantic provinces turned to the college for diagnosis, treatment or necropsy of seals, dolphins, and whales, help for stranded whales, and rehabilitation of injured or sick birds, such as blue herons, northern gannets, and common loons. By the time the college celebrated its tenth birthday, the marine section had come full circle, back to Reg Thomson's vision of the facility. "It all came to pass in the end," Johnson said. "When I said in the beginning that I was only doing farm fish and nothing more, the dean was not a happy camper. We had a long discussion about it, and I convinced him that it was all we could handle with the people we had then. He said, 'I understand that, but we can't lose sight of the fact that we need these other groups.' He lumped them all under marine species — fish, mammals, birds, everything that was marine we were to do, seagulls and everything. And in the end, we did. First we had farm fish, and then we added the others, and it just continued to grow."

Some of the marine species fell under the purview of the Canadian Co-operative Wildlife Health Centre, a network of diagnostic

Pierre-Yves Daoust (right) with female minke whale found dead on the east shore of PEI in October 2003. It may have been bought into shallow waters by the high winds of Hurricane Juan.

laboratories and veterinary colleges across Canada. AVC pathologist Pierre-Yves Daoust, regional co-ordinator of the centre, did hundreds of necropsies of seals and whales through the years, and encouraged the public to bring sick or injured wild birds and animals to the college for diagnosis. The centre was designed to maintain data on the causes of illness and death in wildlife. Animals that had a chance of being rehabilitated were sent to the appropriate clinic, and, if possible, released back into the wild. Faculty and staff did what they could to provide nursing and medical care. Orthopaedic surgeon Dr. Caroline Runyon's caseload included songbirds that had fallen out of the nest, bald eagles with lead poisoning, injured crows, owls, hawks, seagulls, and falcons. "It's not the popular thing to do to fix up crows and seagulls," she observed, "but crows have such phenomenal personalities. Seagulls are kind of grumpy, but we like them, and, if you can fix them, they probably deserve to live just as much as anybody else." Convalescing birds tried out their wings in a flight cage at the college, and, in the case of some bald eagles, in a larger flying school for birds near Montreal.

One patient who didn't make it back home was a snowy owl that

Louie, the Snowy White Owl, with Dr. Caroline Runyon, faculty member in the department of Companion Animals

a conservation officer had brought to the college. By the time the owl was rescued, he had been suffering from a dislocated elbow for some time and was unable to forage for food. He was very thin, and so badly injured, all Runyon could do was to fix his wing so that it wouldn't drag on the ground. Because he could no longer survive in the wild, the hospital obtained permission from wildlife officials to adopt him as a teaching bird. Animal technician Donna Barnes named him Louie. He spent the next six years meditating on his perch, hopping around the small-animal treatment room of the hospital, serving as a demonstration model for classes in restraint and handling, visiting public schools for lessons on birds, and lording it over Basher, the resident cat. "Louie seemed to enjoy it," Runyon said. "The more attention he got, the happier he was. He used to love it when people came for tours. He would fluff himself up and look important." Louie did not welcome everybody to his domain. "He just adored Donna," Runyon said. "He would sit on her shoulder and go through her hair, grooming her. But he hated me, absolutely hated me, because I used to do physiotherapy on him, trim his beak, trim his talons, and make sure his feet were okay. So I was the bad person."

Runyon's caseload also included hordes of homeless cats and dogs. In 1994, the teaching hospital began offering free medical and surgical care, including spaying and neutering, to animals in the local shelter. By 2002, the hospital had neutered more than 950 animals from shelters in Charlottetown, Amherst, and Moncton, and had treated about 750 homeless cats and dogs for sickness and injury.[14] Once well, the animals were returned to the shelters for adoption.

That program was funded by the Sir James Dunn Animal Welfare Centre, which owed its start to Lady Beaverbrook, the widow of two wealthy Canadian industrialists, Sir James Dunn and Lord Beaverbrook. Lady Beaverbrook had a love for animals, especially horses and dogs — she owned a stable of racehorses and a pack of beagles — and, before her death in 1994, she offered to support the college in projects designed specifically to help animals. As a result, the College set up an Animal Welfare Unit in the summer of 1994, with a grant of $125,000, which funded part-time co-ordinator

Basher

He was a tough guy — a big, burly, grey tabby cat that a Good Samaritan had brought to the small-animal hospital shortly after it opened. Nobody knew much about him except that he'd obviously just been in a terrible scrap. He didn't have a name, or, as far as anybody could tell, a home, but he soon acquired both. Marion MacAulay, then manager of animal services, decided he'd make a fine barn cat. The staff called him Basher, after the clinician who had fixed up his wounds when he arrived. That proved to be an apt name. Basher had the run of the place, following MacAulay on her morning rounds every day, through the clinic, up to the front desk, back to the barn. Unfortunately, one day he took offence at the sight of a Doberman in the clinic and sent the poor dog howling to the corner with a big scratch on his nose. After that, Basher was confined to the barn. He grew to despise the clinic, anyway. He had been treated there for a few ailments, and as soon as MacAulay would head in that direction, he'd hiss, turn around, and find his own way back home. Basher eventually became old, incontinent, and too ill to wander about. He died in 1998. But people at the college remembered him for years after that. And for years, MacAulay kept his ashes in her office.

Dr. Alice Crook with Tigger

Dr. Alice Crook's salary and three major projects: the medical and surgical care of Humane Society animals, research on the reduction of post-operative pain in dogs, and research on convalescing equine athletes. Lady Beaverbrook's will specified that the proceeds of a substantial sum from her estate were to be used for animal welfare projects. When the will was probated, the college received a five-year commitment of $2.2 million from the Friends of the Christofor Foundation, created by Lady Beaverbrook's estate. The centre was expanded in 2000, and named after Lady Beaverbrook's first husband, in accordance with her wishes.

Over the next few years, the centre funded a wide range of projects, all aimed at improving the lives of animals, including the subjects of research studies. Almost all of the projects centred on companion animals and horses. In the late 1990s, for example, AVC clinicians began providing medical care for retired racehorses waiting for adoptive homes at the PEI Equine Retirement Society farm in O'Leary, Prince Edward Island. A research project was aimed at developing ways of assessing and managing pain in companion

Open house with Dr. Norma Guy demonstrating "clicker training" of dogs

and wild birds. Veterinary students visited schools to teach children how to take care of their pets and to foster compassion and respect for animals. Students also trained dogs at the Charlottetown animal shelter to eliminate some behavioural problems, thus making them more attractive for adoption.

Lady Beaverbrook's bequest allowed the college to develop an animal welfare program unmatched in size and scope in other institutions, at least in the early years. Crook noted that traditional research-granting sources rarely funded the kinds of projects the centre initiated. "Unlike the research in fish health and farm animals," she said, "the work with companion animals does not attract a great deal of interest and there is not an identifiable group lobbying for funding."

While most of the research at AVC focused on animal health and welfare, faculty and students also took on human and environmental concerns. By 2002, studies dealing with human health included research on cardiovascular disease, diabetes, obesity, and the role of estrogen in post-menopausal women.[15] In the late 1990s, pathologists Pierre-Yves Daoust and Scott McBurney travelled to

Vet Camp

In the summer of 1999, Dean Tim Ogilvie initiated a program to introduce veterinary medicine to students at a much earlier age than usual. In a program unique in Canada, students entering Grades 5, 6, and 7 took part in two week-long day camps at AVC, where, among other activities, they learned about keeping pets healthy, observed surgeries, identified cells and micro-organisms under a microscope, and went on field trips and on rounds in the teaching hospital. Veterinary students served as camp counsellors, and faculty led sessions on topics such as surgery, anatomy, animal welfare, wildlife, aquaculture, and large- and small-animal medicine.

Vet Camp was an immediate hit. Just two weeks after the camp was announced in the first year, all available places were taken. Campers signed up for the experience from throughout the Atlantic region and beyond. From then on, the camp became an annual fixture at the college. "It was an awesome week!" one camper observed as the 2002 Vet Camp ended. "I'll be sad to go back home."

hunting camps in Labrador, at the request of the Innu Nation, to try to determine whether contaminants such as PCBs, dioxins, residues of DDT, and heavy metals were affecting the health of food sources such as caribou, porcupine, and geese — the ultimate object being to determine the effect, if any, on human health.[16] In a research project closer to home, Dr. John VanLeeuwen was determining the geographical distribution of nitrate contamination in drinking wa-

Open house at AVC

ter in Prince Edward Island, and the possible relationship between nitrates and the incidence of Type I diabetes. Dr. Alastair Cribb, a specialist in human and veterinary pharmacology and director of the world's first Laboratory of Veterinary Pharmacogenetics, was investigating, among other things, the role of genetic makeup in the effects of drugs.

Projects such as these enhanced AVC's image in the eyes of the broader community and underlined its relevance to the people whose tax dollars were helping to keep it going. As UPEI President Wade MacLauchlan observed, AVC research, being mainly applied research, was seen as inherently useful. "And it's also work that communicates well," he said. "One of the things I find as president is that people will at this point speak with some enthusiasm and recognition to me about the work that is done at the Veterinary College."

For some people in the community, AVC was just a place where young people went to study veterinary medicine. For others, it was a neighbourhood animal hospital. Some were aware of, and affected by, the many projects initiated by the college in the animal welfare

Dr. Norma Guy and Tek unveil the plaque recognizing the contribution of the Friends of the Christofor Foundation and the Sir James Dunn Foundation for the establishment of the Sir James Dunn Animal Welfare Centre (September 27, 2000)

field. Some recognized the value to society of research taking place at the college. Probably few realized the complexity of AVC's outreach to the community as it tried to fulfill its mandate of teaching, service, and research.

When Peter Meincke was president of UPEI in the 1970s and early 1980s, one of his fondest hopes had been to bring the university closer to the surrounding community. He may not have seen that dream realized during his tenure, but he did help it come to pass eventually. In partnership with Reg Thomson, Meincke worked hard to bring the proposed new veterinary college to Prince Edward Island. He believed that, in addition to bringing more education dollars to the Island, the college would strengthen the university and enhance its image. Two decades later, he concluded that AVC had far exceeded his and Thomson's dreams. "My impression is that it has had a really good impact on the university," he said. "And, most of all, I get the really strong sense from people in the community that everybody is glad it's here. Very glad it's here."

10: Making an Impact

In the beginning, not everyone was glad to see the new veterinary college. Before it was built, some veterinarians, including the executive of the Canadian Veterinary Medical Association, argued that, if there was a shortage of veterinarians — and some had doubts about that — any manpower needs could be filled, at less expense, by sending students to one of the existing North American colleges. Some Maritime veterinarians feared that there simply wouldn't be enough work for all the new practitioners the college would be churning out.

Fifteen years later, that fear was not borne out. The graduates were each getting a couple of good job offers, if not always within the Atlantic region. Dr. George Irving, a small-animal practitioner in Moncton, New Brunswick, recalled that, before the Saskatoon college opened in 1969, similar anxieties dogged the profession. "Oh, people were moaning and groaning," he said. "There would be veterinarians on the street looking for jobs. Well, they put a veterinary college in Western Canada and you'd never know it was there. They filled it with people from Western Canada, and they all got jobs. Then we talked about building a veterinary college in Atlantic Canada. It was the same story. 'Oh, what will we do with the veterinarians? Where will they go?' Well, there are no veterinarians at the unemployment office. Everybody got work. And there's still a shortage of veterinarians."

On the Island, where the veterinary caseload was limited by geography, one fear was that the new college would lure clients away from existing clinics. Dr. Claudia Lister's small-animal practice in Stratford was just across the river from the university campus — a little too close for comfort, in her view. A lot of the growth in her practice came from pet owners seeking a second opinion on a case. When the teaching hospital first opened, she said, some of the "second-opinion" business went to the AVC clinic instead, because

Miss Hagerman's Gift

Verna B. Hagerman was born and raised on a farm at Bear Island, near Woodstock, New Brunswick, where she apparently developed a love and concern for animals. After obtaining a master's degree in English from McGill University, she taught school for thirty years

When she was about eighty-five, she wrote to the Holland College vocational and technical school in Charlottetown, offering to leave a sum of money in her will for a student in the new veterinary school. The principal of the day redirected her to the University of Prince Edward Island, and, in February 1985, the year before the veterinary college opened, she wrote to AVC Dean Reg Thomson. "Barring a prolonged terminal illness in a nursing home," she wrote, "I should leave some money, which I should like to be used as a bequest to a course in Veterinary Medicine you propose to establish in 1986. If acceptable to you, the bequest would be to help train a worthy young person for a career in animal care." Thanking her for that offer, Thomson suggested that she consider naming the scholarship "The Verna Hagerman Award in Veterinary Medicine."

Miss Hagerman died in November 1991, at the age of ninety-one. In her will, she left a bequest of $645,000, to be used to provide scholarships for veterinary students of ability. The awards, named "The G. Murray and Hazel Hagerman Scholarships" in memory of Miss Hagerman's late father and sister, amount to about $3,800 each. They are given every year to nine students who stand first, second, and third in their classes in each of the first three years of the DVM program. In addition, three Hagerman scholarships, each amounting to one year's full tuition, are given each year to graduate students.

it was widely assumed the hospital could offer better emergency service and lower fees. As people realized that all clinics were obliged to offer emergency services, and that private clinics had some advantages that the hospital lacked, her caseload stabilized.

In Lister's view, one advantage private clinics had to offer was that of accessibility: At the teaching hospital, clinicians had to divide their time among teaching, research, and clinical duties, with the result that clients were not always able to speak directly to the veterinarian in charge of a case. On the other hand, the AVC clinic offered specialized procedures, such as total hip replacements, not available elsewhere in the region. Before the AVC hospital opened, Lister had referred a few — but very few — special cases to the Ontario Veterinary College. "I think you have to give the college a lot of credit as far as bringing professional resources to this area," she said. "It's now made it more possible to take veterinary medicine to that higher level. We don't do brain surgery here, or spinal surgery, or total hip replacements."

H. Wade MacLauchlan, fifth president of UPEI

Dr. Andrew Peacock, employed by the provincial government in a one-man, large-animal practice in rural Newfoundland, found that he relied on phone consultations with faculty at AVC. "I work alone," he said, "so if I need advice on a case, I either phone another practice or AVC. In fact, I spent half an hour yesterday with a parasitologist at the college." Dr. Carl Dingee, a large-animal veterinarian working out of Moncton, was using the college "routinely" for both referrals and phone consultations.

Dingee benefited from the college in yet another way: he was one of hundreds of young Atlantic Canadians who graduated from veterinary school without leaving the region. In its first thirteen years, 524 Atlantic Canadian students, 102 international students, and 21 Canadian students from outside the region graduated from AVC with doctor of veterinary medicine degrees; 110 students obtained master's degrees from the college, and seven students went through the PhD program, established in 1996. By the year 2002, Prince Edward Island had seats for 10 first-year students in the DVM program; Nova Scotia had 16; New Brunswick, 13; and Newfoundland, 2. In addition, 19 seats were sold to international students, who were paying $42,076 a year in tuition, compared with $7,100 for students from the Atlantic region. As of June 1999, about 35 per cent of the 591 practising veterinarians in the region were AVC graduates.

As the college entered its sixteenth year, Dr. Ross Ainslie of Halifax, one of the most prominent and persistent critics of the college, was still arguing that the Atlantic region's veterinary needs could have been served more cheaply by expanding classes at the Ontario Veterinary College. And if a fourth college had to come to the region, he still would have preferred to see it at the Nova Scotia Agricultural College in Truro. Still, like many another veterinarians in the region, Ainslie had become accustomed to AVC. So much so, he said, that he didn't know how people got along without it. "You can call up and discuss cases [with the college faculty]," he said. "It doesn't mean they'll only deal with us if we send clients over there. They do lab diagnostic services as well. It's very useful. The younger generation, especially, would not be able to practise without it."

Not surprisingly, the work of the college also shone a favourable light on the university of which it is a faculty. Wade MacLauchlan, appointed president of the university in 1999, said the faculty of Veterinary Medicine had helped make the university more relevant in the eyes of the wider community, partly because of the nature of AVC's research: before the college came on the scene, the university was known as an undergraduate, liberal arts, teaching institution. After the college opened, the university established graduate programs and gradually acquired the facilities and funding it needed to develop a significant research role. With the retirement of some old hands, it also acquired new blood, eager to embark on research projects. By 2002, the Science faculty was pulling in as many research dollars as the veterinary faculty. By 2003, the university was expected to have the fourth-highest level of peer-reviewed research in the Atlantic region. "I think the veterinary college certainly helped to set the standard," MacLauchlan said. "What I'm really proud to say is that this is an institution —they're very rare — that has really moved forward into a research-intensive world while still keeping teaching excellence as a central value of the institution. I think the vet college has enhanced that."

In terms of dollars and cents, Prince Edward Island, the host province, was the most obvious beneficiary of the existence of the college. A 1995 study by University of Prince Edward Island economics professor Annie Spears[1] estimated that, through its expenditures and those of faculty, staff, students, and visitors, AVC generated about $40 million and created about 325 jobs annually on the Island, in addition to the 250 it provided directly as an employer. Total tax revenues accruing to the province as a result of income generated by the operation of AVC amounted to more than $7 million annually, for an initial investment of $5 million (in grants) by the province.[2]

But, as the Atlantic Provinces Economic Council (APEC) observed in an October 2000 report on the economic impact of AVC on the region, the other three provinces also appeared to be getting more than their money's worth from the college. While noting the difficulty of quantifying the benefits of much of the work carried out by the college, the report concluded: "The provincial funding received by the college seems quite small in relation to the potential economic impact of AVC's teaching, research, and service activities." Even without taking into account the value of a regional veterinary college for students, and the contribution the college makes to pet and equine industries, human health, and the environment, the APEC report estimated that AVC's activities in the fields of teaching, research, and service would need to contribute less than one per cent to the income or output of the Atlantic region's livestock, aquaculture, and lobster industries to justify the provincial funding. "AVC's contribution and impact on the livestock, aquaculture and lobster industries is likely to be more than sufficient to justify its provincial funding," the report concluded. "Expenditure on veterinary services by livestock farmers is about 2.3 per cent of farm revenue. AVC educates the veterinarians that help look after the livestock in these industries. It provides health and productivity information services, diagnostic services, specialized treatment and surgery, and expert advice that veterinarians rely on to conduct their work. Moreover, AVC researchers are investigating a number of diseases that can cause losses of 10 per cent or more of the fish or livestock .... Post-harvest losses in the lobster industry, for example, amount to 10 to 15 per cent, and researchers at the Lobster Science Centre are investigating the cause of this loss in order to help reduce it. Other research is investigating or testing new treatments, new feeds or supplements, new technology, or new husbandry practices that will help reduce costs or increase productivities in these industries. The potential impact of AVC's activities on these industries is therefore quite significant."[3]

Under the interprovincial funding agreement, by 2001, Prince Edward Island was funding 38 per cent of operating costs, or $4,466,840; Nova Scotia was contributing 32 per cent, or $3,767,350; New Brunswick, 26 per cent, or $3,061,000; and Newfoundland, 4 per cent, or $471,050. In addition, Prince Edward Island provided $450,585 to support the graduate program, while the other three provinces contributed a total of $40,800 to provide graduate student stipends. In the late 1990s, when governments were hacking away at publicly funded programs to reduce deficits, the provincial governments asked that operating grants, as a percentage of total revenues, be reduced from 69 per cent to 60 per cent during the five years ending in 2000–01. The AVC business plan set an objective of 57 per cent, but the college in fact managed to surpass that. In 2000–01, revenues generated by the college, mostly through tuition fees, sales and service fees, and research income, accounted for 49 per cent of its operating budget,[4] requiring the provinces to contribute only 51 per cent.

Still, that did not settle the funding issue. AVC was a child of the region, but the region was not always one big happy family. There still existed remnants of political antagonism that had marked the debate on the birth of the college three decades earlier. That became clear every time a new interprovincial funding agreement to operate the college was discussed. For instance, the 2001–06 agreement was to have taken effect in May 2001; it was not signed until May 2003. Nova Scotia, indicating during the discussions that it could not at that time commit to its usual financial support, took advantage of a provision in the agreement that allowed it to ask the college to market some of its seats to international applicants. Nova Scotia chose to give up five of its sixteen seats for the last four years of the agreement. That meant that, at least until 2006, the number of international students rose to twenty-four per class.

Despite such setbacks, the college continued to set ever-higher

Dean Tim Ogilvie

Dr. Timothy H. Ogilvie, the third full-term dean at AVC, was one of the first faculty members recruited by the college before it opened in 1986.

Ogilvie came to the college with experience in teaching, research, clinical practice, and government service. Born in Fort Erie, Ontario, he spent summers as a youth on his grandparents' farm, and enrolled in the Ontario Veterinary College in 1971. After graduating in 1975, he took a job at a mixed practice in Kensington, Prince Edward Island, for two years. During that time, he fell in love with the Island and with his future wife, Lola Meek. However, he left his adopted home long enough to complete a master of science degree and then to teach at the Ontario college for three years. Returning to the Island in 1982, he worked as Provincial Veterinarian before being recruited as a professor in large-animal medicine in AVC's department of Health Management. He became acting chair of the department in 1990 and chair in 1991, was appointed acting dean of the faculty of Veterinary Medicine in 1998, and full-term dean in 1999.

In Ogilvie's first term as dean, he oversaw a period of growth in research, teaching, and service at the college. One of his priorities was to work in concert with his counterparts at the three other Canadian veterinary schools to obtain funding to expand and renovate facilities. The goal of this partnership was to maintain accreditation at the colleges and increase their capacity to undertake research in such areas as animal health, animal welfare, public health, food safety, sustainability of production systems, and the safeguarding of trade. "It was a very vulnerable time for our four colleges," Ogilvie said, "in that two were facing difficulties maintaining their accreditation. Therefore, it has been a time of significant challenges for us, but, at the same time, we have opportunities for better linkages of the four colleges and a chance to step up to the plate and do something great for Canada." As a result of this initiative, the federal government announced in December 2002 that, as part of a Canada-wide project, AVC would receive $18 million for its upgrading and expansion projects.

Ogilvie's other leadership roles included terms as chairman of the board for AVC Inc., chair of the board for Genome Atlantic, president of the Canadian Faculties of Agriculture and Veterinary Medicine, and executive board member of the Association of American Veterinary Medical Colleges. In 1991–92, he also served as president of the Canadian Veterinary Medical Association.

H. Wade MacLauchlan, president of the University of Prince Edward Island, described Ogilvie as "a model dean" with uncommon leadership qualities. "He is brave and entrepreneurial, fair and patient. He's a good person — and that is a very important part of having a successful institution, having someone that cares and dares and laughs and sees the quirks, and then works with others in a very sustained, mainly quiet, but very efficient way."

goals in teaching, research, and service. In 2002, Dr. Tim Ogilvie, who became acting dean in 1998 and full-term dean a year later, was working in concert with deans of the other three veterinary colleges in a request for federal government funds to renovate or expand the four schools. AVC's proposal included a plan for about 60,000 extra square feet of research and office space, as well as renovations to the hospital, aquatic-animal facility, and diagnostic areas. In addition, the plan included "biosecurity level-three" laboratories at the schools, which would be able to identify and monitor contagious-disease organisms that regular provincial animal-pathology labs were not equipped to handle. This would give the schools greater capability for bio-surveillance, professional training, and emergency response to such diseases as West Nile Virus and Bovine Spongiform Encephalopathy (Mad Cow Disease).

AVC, unlike the other colleges, was not at that point facing deterioration of its infrastructure. What it desperately needed, though, was more space. In 1996, when the class size increased to sixty, modest renovations were made to the college to accommodate the additional students. As the numbers of faculty, technicians, and graduate students increased toward the end of the decade, and opportunities developed to attract more research dollars, overcrowding became even more acute. "We've got graduate students hanging from the rafters, and we've got faculty who don't have labs to conduct their research in," Ogilvie said. "My message is, we could do so much more for the needs of the nation in terms of research, prevention of zoonotic diseases, food safety, public health, animal health — all those things. We could compete very well, but we need more space to do it." That plan came closer to reality in the winter of 2002–03, when Ottawa announced that, as part of a four-college funding program, AVC would receive $18 million for its planned expansion.

Ogilvie also foresaw that AVC could be a catalyst for new, related industries in the community. That had been a goal of former premier Jim Lee, and it started to happen in the early years of the 21st century, with such spin-offs the establishment of a company by an existing private-sector firm eager to commercialize technologies produced by the Lobster Science Centre. In addition, by the fall of 2002, plans were under way for a major health research centre on the UPEI campus in which the college would be a partner. The proposed

National Research Council Institute for Nutrisciences and Health would, among other things, identify, purify, and synthesize bioactive substances from aquatic- and land-based natural resources such as blueberries, which might be useful in treating health problems such as neurodegenerative disorders, cardiovascular disease, or obesity.

In the early 1970s, at the beginning of the long and arduous gestation period that preceded AVC's birth, some people envisioned that a new college would benefit the region's agriculture industry; a few saw help for the burgeoning aquaculture industry; some focused on educational opportunities for students and practising veterinarians; some foresaw an impact — for good or ill — on the host university. It is doubtful that anyone could have imagined the depth and breadth of the influence the college would have on the Atlantic region through the teaching hospital, the Fish Health section, the herd-health programs, the diagnostic facilities, the animal welfare projects and the wide-ranging research studies carried out by the college's four departments. Those activities created a reputation for the college throughout the region and, in some cases, around the world. And, as AVC entered its seventeenth year, it looked as though that influence and reputation were destined to continue to grow.

11: Plus Ça Change...

When Dr. Sylvia Craig was a student at the Atlantic Veterinary College, a professor asked her one day to administer oral medication to a cow. This did not go as planned. "I had my arm around the cow's neck," Craig recalled, "and she lifted her head up, and both my feet came off the ground, and I thought, 'Well, maybe I should really think about doing something else.'" Years later, she remembered that as a defining moment in her career path. Although Craig had grown up in Halifax, she'd always been interested in large animals, especially horses. She didn't even mind the fact that, at five-foot-two, she had to stand on a hay bale to do rectal palpations on cattle during large-animal rotations at AVC. Until that cow-induced epiphany, she was still considering large-animal medicine. "It could have been done," she said, "but I would have had to be constantly proving myself."

Craig graduated at the top of her class in 1990, took a job for a year in a small-animal practice, and then switched to a laboratory at Dalhousie University in Halifax. With two young children, she had found the clinical work a struggle. "It was quite difficult to manage with a young family," she said, "especially when I was expected to work evenings. I'd go in at supper time and work until eight or nine o'clock, and it meant I never saw my kids."

In some respects, Craig was fairly typical of the young people entering the veterinary profession around the turn of the 21st century — predominantly female, urban, and oriented toward either a small-animal practice or a job that didn't demand attention nights and weekends. The makeup of the profession had changed dramatically from the 1970s, when discussions about a fourth veterinary college had begun. Some of the issues had not changed — low incomes, long hours, a shortage of veterinarians in some fields, intractable diseases — but some aspects of the profession had been turned upside-down. There were greater demands for specialization and for competence

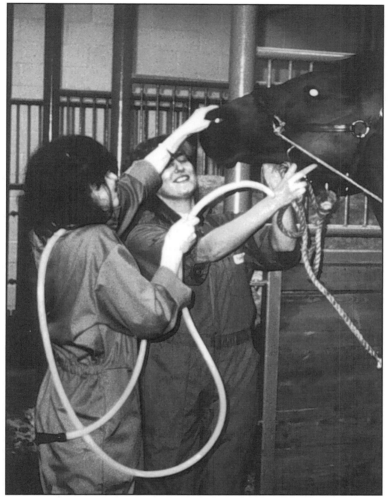

Carol Brown and Jackie Rideout, students in the Class of 1990 (Fall 1986)

in new roles in corporate agriculture, in aquaculture, and in the increasingly complex areas of environmental and public health. A more knowledgeable companion-animal clientele was demanding an ever-higher level of care. And the workplace was no longer a male-dominated territory.

When Dr. George Irving of Irishtown, New Brunswick, enrolled in the Ontario Veterinary College in the mid-1950s, five of the sixty-

Dean T. H. "Tim" Ogilvie (on right) presenting Mr. Wu Yuanyuan, head of a visiting Chinese delegation, with a copy of the AVC Lindee Climo poster (original Climo painting *Night Group* is on wall behind them)

five students in his class were women. In 1976, when Dr. Claudia Lister graduated, about 25 per cent of the class were women. The first graduating class at AVC was equally divided between men and women. And, by 2001, women comprised 80 per cent of the AVC class. That trend was consistent with what was happening elsewhere on the North American continent. According to a 1999 study, women made up nearly 70 per cent of veterinary students in the United States and 36 per cent of practising veterinarians. The authors of the study predicted that the female proportion of veterinarians would increase to 50 per cent by 2004 and 67 per cent by 2015.[1]

A number of reasons have been advanced for this change. One is the rise of the women's movement and its demand for equal access to the workplace in general. Another is the establishment of new veterinary colleges in Canada and the United States, which opened the doors to more students of both genders. The image of the profession also changed. As society became more urbanized and pets became more and more like family, companion-animal practices sprang up like mushrooms. As large industrial farms replaced small family-owned operations, and herds became healthier because of preventive medicine practices, large-animal work declined. As a result, the profession took on a more compassionate, less utilitarian image — one that attracted many women. "Frankly," said AVC pathologist Pierre-Yves Daoust, "women tend to be more compassionate about other living beings than men, although there are exceptions." Besides, working indoors with small animals is generally less physically demanding than barnyard medicine, although physical size and strength have become less important with the development of safer tranquilizers and sedatives.[2]

Academic achievement has been another factor in the large enrolment of female students. At AVC, most students accepted into the program come equipped with marks in the mid-8os. Dr. Graham Jones, a small-animal veterinarian in St. John's, Newfoundland, observed, "Girls tend to have fewer distractions, and perhaps they're willing to work harder."

Money — or the lack thereof — is another reason some people give for the feminization of the profession. Veterinarians generally earn less than other professionals with a similar amount of training and a similar workload. Some people speculate that women veterinarians are more likely than men to accept relatively low salaries — partly, no doubt, because women tend to undervalue their skills, and partly because some expect their partners will be the primary wage-earners. "If men want to make money, they become doctors or dentists," observed Dr. Natasha Cairns, a small-animal veterinarian in Windsor, Nova Scotia. The average top-level salary reported for the veterinary profession in Canada in 1997 was $60,000, compared with $122,100 for dentists, $114,500 for physicians, and $68,400 for pharmacists.[3] At the same time, veterinary student debt loads were climbing. A 1996 survey[4] showed that graduates from the three English-language veterinary schools reported an average debt of $28,600; Saint-Hyacinthe graduates had an average debt load of $15,000. A 1997 study indicated that only 24 per cent of new graduates from Canadian colleges made more than $40,000 a year.[5] And female veterinarians were making even less than their male counterparts. One American study reported that male graduates could expect more job offers, larger salaries, and, usually, larger benefit packages than female graduates applying for full-time employment.[6] (Another study noted that this was not a condition specific to veterinary medicine; in fact, the wage gap was wider in fields such as dentistry and human medicine.[7]) To compound the income difference, many women worked fewer hours than men because of child-care responsibilities.

Women are also less likely than men to own their own practices. Natasha Cairns, who graduated from AVC in 2001, felt that she could be making more as an owner than as a clinic employee, but had "not one inkling of interest" in setting up her own practice. "Owners of clinics sometimes lose touch with why they got into veterinary medicine in the first place," she said. "I do not want to become like that, where it becomes a business rather than a service." Besides, she planned to start a family soon, and hoped to work only part-time while her children were small.

Dr. Sherri Coldwell, a 1991 AVC graduate, discovered how difficult it is to combine clinical work with family responsibilities. In the summer of 2002, she was employed in a small-animal clinic in Yarmouth, which meant a schedule of 8 A.M. to 5 P.M. five days a week, working every second Saturday morning, and being on call two nights a week and every second or third weekend. Besides this, she was practising equine medicine in her spare time. That left her little time to spend with her husband and seven-year-old son and, she said, created a strain on her family life.

The conflict between home and career responsibilities also causes problems for clinic owners. Dr. Claudia Lister of Stratford, Prince Edward Island, said a number of women veterinarians had applied for part-time work at her small-animal clinic. "They don't want to work on call or work nights or weekends. We need to have someone to respond to emergency calls, so if we have an associate who doesn't want to be on call at all, it makes it very difficult for the rest of us."

On the other hand, women have helped change the image of the profession in a very positive way. If women were being attracted to veterinary medicine because of its increasingly compassionate image, they were also contributing to that image. Dr. Caroline Runyon, an orthopaedic surgeon at AVC, said some clients reported finding women veterinarians more understanding, more patient, and more willing to discuss a case in detail.

That kind of attention fits neatly with changes taking place in society's view of animals, especially companion animals. As AVC Dean Tim Ogilvie observed, "There are a lot of similarities between paediatrics and companion-animal veterinary medicine." Farmers, however much they love their cows, generally are obliged to measure their worth in monetary terms; many urbanized owners of companion animals see their pets as priceless members of the family. Dogs and cats, once relegated to the barnyard, now sleep on the bed, travel in the family car, have their teeth brushed, wear sweaters and booties, are buried in pet cemeteries. Animal welfare and environmental groups have helped raise the status of animals in the eyes of society. So has public awareness of the value of animals as therapists and helpers for the disabled. "People have started to realize that animals are sentient living things," Daoust said. "They feel stress, physical pain, mental stress. As this has progressed, the veterinary profession has been the logical one to be asked, 'Well, what do you think of the well-being of these animals?'"

The changes in attitudes toward animals are reflected in veterinary education. Runyon, a graduate of Colorado State University, recalled that she was taught little or nothing at school about pain management. "They talked about surgery, but never what to do after the surgery. There is an incredible emphasis now, and I think a good emphasis, on managing discomfort and trying to find out the things dogs and cats do that let you know they have pain."

One of Runyon's worst memories of veterinary school was surgery lab, where seven or eight surgical procedures were performed on one dog before it was finally euthanized. "We didn't like it, but we didn't have any choice. It was just horrible, because you were emotionally attached to this dog by the end of the semester and then you had to kill it. That doesn't happen any more. We do cadaver surgeries on animals euthanized for other reasons at the Humane Society. All our live surgeries are spays and neuters of dogs and cats. That has changed dramatically and, I think, for the good."

Daoust, a 1974 graduate of the Université de Montréal faculty of veterinary medicine at Saint-Hyacinthe, could not recall that his classmates had ever discussed issues such as the quality of housing for the school's teaching animals. Thirty years later, students at veterinary schools were regularly questioning how animals were being used for teaching and research. At AVC, strict rules were in place governing the use of teaching animals: a cow subjected to a rectal examination, for example, would have to be given a rest period before being disturbed for any reason. Even so, the Class of 2001 set up a fund to improve the social and environmental lives of the teaching animals. "There is complete acceptance by everybody in here," hospital director Dr. Wendell Grasse said, "that animals deserve the same level of care and comfort and attention as we do." Cosmetic surgery such as tail-docking and ear-cropping, while not explicitly prohibited by the teaching hospital, was left to the conscience of individual clinicians, and rarely, if ever, performed. At the same time, the college was leading the way, primarily through its Sir James Dunn Animal Welfare Centre, in teaching, research, and service activities aimed specifically at improving the welfare of animals.

An obvious corollary to the human-animal bond is that pet owners demand a high level of medical care — and are willing to pay for it. "In veterinary private practice," the authors of an American veterinary market study observed, "recognition of the human-animal bond is an important determinant of a successful practice. There is a growing recognition that provision of veterinary services in a manner that acknowledges the human-animal bond will lead to better outcomes for veterinary practices and their patients." With more disposable income than was available a generation or so ago, clients are more able to pay for veterinary care. As Dr. Andrew Peacock of Carbonear, Newfoundland, expressed it: "Twenty years ago, if a cat needed $500 worth of work, people would say they couldn't afford to fix the cat. Today, there's no end to the money some people will put into treatment for their pets. There is a demand for sophisticated treatment, and they will pay for it if you can provide it."

Some clients, caught up in the prevailing rage for unconventional therapies, are also looking for the same alternatives to prescription drugs and surgery that they seek for themselves — herbal remedies, supplements, homeopathy, acupuncture, touch therapy. In 1998, a CVMA task force on the future of the profession noted that consumers were increasingly looking for alternative treatments. And, although not all veterinarians should be expected to use unconventional therapies, all veterinarians should at least be familiar with the rationale for their use. If not, the task force warned — resurrecting a spectre from half a century earlier — non-veterinarians would be willing to step in. "The encroachment of lay people ready and willing to practise veterinary medicine will be a growing problem for the profession," the authors predicted.

By the time that report was published, AVC had already dipped a toe into the alternative-medicine pool. In 1996, equine surgeon Dr. Art Ortenburger began a clinical acupuncture service at the college. Ortenburger, originally skeptical about the value of non-Western medicine, had taken a veterinary acupuncture course the previous year in hopes that it would provide an answer for intractable back pain in standardbred horses. Because of the nature of the caseload at the teaching hospital, he ended up treating more dogs and cats than horses, some of his notable successes being with back pain, heaves, and hip dysplasia. He also taught a course in acupuncture to fourth-year students, and found that some graduates used that training as a negotiating tool when taking a job. "If you look through the want ads in back of the veterinary journals," he said, " there's a new kind of ad. In almost every issue there's a practice somewhere looking for somebody to do homeopathy or chiropractic or acupuncture."

Because of a lack of a good body of scientific research to validate alternative therapies, Dean Ogilvie said, veterinarians would need to acquire expertise in this area through self-study. Alternative methods would not likely be included in the core undergraduate curriculum for some time. What he did foresee was a focus on other programs that would promote the diversity of the veterinary profession. "In the past, we have had veterinarians serve society in many ways," he said, "including meat inspection, public health, human-health research, laboratory animals, and policy and planning for agriculture. I would like to see veterinary medicine make inroads into areas where traditionally we have been strong but lately have not stepped up to the plate to take an active role." This could include food and water safety, research into drugs for both animals and humans, and investigation of new and re-emerging diseases such as West Nile virus and tuberculosis.

In the first fifteen years of AVC's existence, most of its graduates appeared to be following the well-trod path of clinical practice. Surveys of graduates indicated that, a year after graduation, 80 to 93 per

Enjoying a lighter moment at an AVC ice cream social (*left to right*): Mel Gallant, Dr. R. A. "Bob" Curtis, Dr. T. H. "Tim" Ogilvie, Dr. L. E. "Larry" Heider

Dr. Ole Nielsen, former dean of OVC and of WCVM, with Dr. Bob Curtis, chair of Health Management. Dr. Nielsen was the first Dr. R. G. Thomson Invited Lecturer at AVC, in 1995.

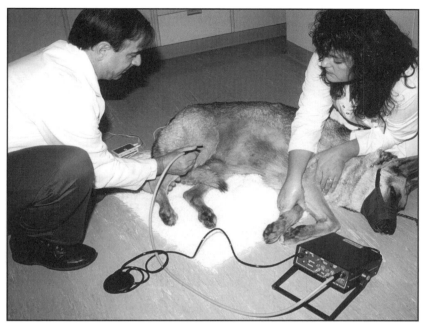

Dr. Art Ortenburger, Large Animal Surgeon, applying acupuncture (with assistance of Animal Health Technician Heather MacSwain)

cent were in clinical practices outside a university, and most worked in practices that were exclusively or predominantly small-animal clinics (64 per cent in 1990; 77 per cent by 2001). Of those from the classes of 1999 and 2001 who responded to the survey, none reported working in a practice that dealt exclusively with food animals.

The diminishing numbers of students interested in large-animal work has worried some members of the profession. In a lecture in 2001 at the Western College of Veterinary Medicine, Dr. Ole Nielsen, Professor Emeritus at Guelph University, warned that this had become a critical issue. He added, "One can also point to the new graduate's relative disinterest in public health, ecosystem health, laboratory investigation, and research. All these areas are vital to our society. We neglect them at our peril." Dr. Bob Curtis, formerly chair of the department of Health Management at AVC, pointed to the outbreak of foot-and-mouth disease in Saskatchewan cattle in the early 1950s. "I remember what a devastation that was, not only

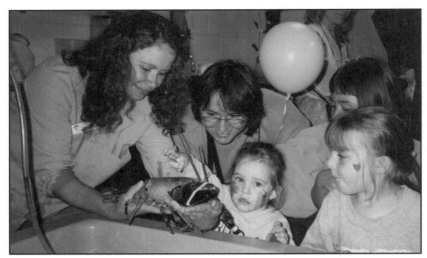

AVC open house

Mel Gallant, assistant dean, Administration and Finance, administering treats to Cairn Terriers Rufus and Max

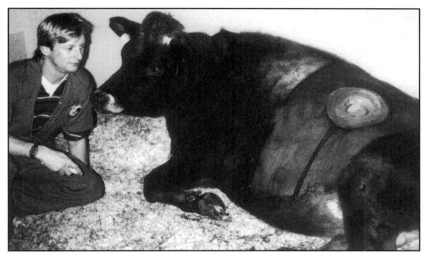

Bill MacDonald, student in the Class of 1990

to farmers but to the whole country," he said. "I'm a little afraid now that if we don't do something more to support large-animal veterinarians, we are going to lose them. And when we have something like foot-and-mouth or some other disease, it could be there for months before anybody's going to know it."

The relative lack of interest in large-animal medicine undoubtedly was related to the decline in the numbers of veterinary students with farm backgrounds. Another factor, according to Dr. Larry Heider, AVC dean through most of the 1990s, was the decline in male applicants for veterinary school. "Where are the young men?" he asked. "We know women contribute to the profession equally as men do, with perhaps one notable exception. And that is, although they come to school and get out of college with an interest in food animals, the statistics say they don't stay in food-animal practice in the same numbers that men do." When he graduated from Ohio State University in 1964, the profession had a more utilitarian aspect. "It was a more robust, rigorous profession than today, certainly in food-animal practice. I think some of us were attracted to that. It was a little more adventurous. You were rousting about the countryside — that James Herriot type of thing."

Some have argued that, to attract the kinds of young people the profession needs, the selection process in veterinary schools has to change. The CVMA's 1998 task force on the future of veterinary medicine predicted a surplus of veterinarians trained for clinical work, especially companion-animal practice, and a deficit in non-practice professionals. Instead of placing so much emphasis on marks, the task force suggested, colleges should try to select people with broader interests than traditional practice and give more weight to such traits as people and communication skills, critical thinking, and compassion for people as well as for animals.[8]

Dr. Graham Jones of St. John's said selecting high academic achievers does not necessarily produce good practitioners. "You tend to get an introverted type of person," he said. "Perhaps they're smart because they're willing to regurgitate ideas back to the professor without having many original ideas." Dr. Andrew Peacock of Carbonear said the selection criteria eliminate some worthy candidates from his province, which was funding only two seats for students each year at AVC. "I have kids going on my rounds with me that I know could handle the academic work [in veterinary school]," he said, "and they love the work, they're just obsessed with it, but they won't even be able to get an interview, and it just breaks my heart."

In the early 1970s, when discussions began on the need for a fourth veterinary college in Canada, there were two main arguments in favour of building one. One pointed to the fact that so many bright young Atlantic Canadians were being deprived of the opportunity to get a veterinary education. Another held that there was a chronic shortage of veterinarians, which would continue into the future. Not surprisingly, thirty years later, there were still disappointed students in the region. And the debate over employment needs had not been settled. In 1998, the CVMA manpower study reported that there appeared to be an oversupply of veterinarians in Canada, particularly in companion animal practices in urban areas, and particularly in the Maritimes and Quebec, although some sparsely populated rural areas suffered a chronic shortage of practitioners. The oversupply problem would continue and drive salaries further downward if current graduation levels were maintained, the task force warned.[9]

However, there always have been divergent views on manpower forecasts, including the fundamental question of whether they are at all reliable. The CVMA was making pessimistic prognostications in the late 1970s, when it opposed the construction of the fourth veterinary school. More than two decades later, graduates from that school were still getting at least two good job offers each. In Dean Ogilvie's opinion, supply/demand analyses require taking into account factors that are almost impossible to predict, including the future economic health of the nation, the success of job-creation programs, and human and pet population trends. In the early years of the 21st century, he pointed out, there was already a shortage of some professionals, with the likelihood of a greater demand by 2015. That meant that, as early as 2003, AVC needed to get high school students thinking about a career in veterinary medicine.

Students about to graduate in 2003 faced, in some respects, quite a different workplace from the one Prince Edward Island's Dr. Bud Ings or Nova Scotia's Dr. Ross Ainslie encountered fifty years earlier. Small-animal clinics, once a rarity, were in the majority. Women practitioners, once few and far between, comprised about half of the profession. Huge advances had been made in diagnostic and therapeutic procedures. Many clients demanded the same level of care for their pets as they would for their children. Farms were fewer, larger, and more complex. Preventive health care had become more valued. Attitudes of younger practitioners had changed; many no longer accepted the prospect of a lifestyle that required being on call twenty-four hours a day, seven days a week.

And yet ... some things had hardly changed at all. The focus of the profession was still on service — to animals, and to humanity. Idealistic students and newly hatched veterinarians still ranked income low on their lists of priorities — and older veterinarians, facing the cold realities of bill-paying, still complained about too much work for too little pay.[10] Manpower forecasts were still regarded with some suspicion. Lay veterinarians were once again hovering in the wings. Practitioners still had to deal with intractable ailments, unreasonable clients, and unpredictable patients. To many of these clients and patients, the family vet was still a hero. And there was still an element of joy in the day-to-day struggle. Andrew Peacock, formerly from Northern Ontario, graduated from the Ontario Veterinary College in 1982, and went directly to Newfoundland, where he is employed by the province in a solo large-animal practice on the Avalon Peninsula. It's a job straight out of a James Herriot story

Dr. Andrew Peacock with an orphaned moose, brought to the Salmonier Nature Park on the Avalon Peninsula in Newfoundland and Labrador

— driving for hours through rain, sleet, and snow, reaching for the phone in the middle of the night, slip-sliding around muddy barnyards to treat colicky horses and castrate rebellious bulls. He loves it. "When I was looking for a job," he recalled, "I wanted one with both mental and physical challenges. When I was interviewed for vet school, one of the veterinarians on the committee told me there was no romance in veterinary medicine. I've been at it over twenty years now — and I still think he's wrong."

Appendices

APPENDIX 1

Chronology

1969 — The University of Prince Edward Island is incorporated. Dr. Ron Baker becomes its first president.

1969 — The Canadian Agricultural Services Co-ordinating Committee (CASCC) decides more information is needed on the veterinary manpower situation.

1970 — The federal government commissions a study to assess the need for veterinarians.

1971 — The CASCC determines that consideration should be given to a fourth veterinary college.

1975 — Maritime Provinces Higher Education Commission (MPHEC) appoints Dr. Dennis Howell of the Ontario Veterinary College to study the need for, and feasibility of, a veterinary school in the Atlantic region.

1975 — Howell recommends the University of Prince Edward Island as the location. This is endorsed by the MPHEC, but the Council of Maritime Premiers cannot reach an agreement on the site.

1978 — PEI government announces it will build a veterinary college with or without the other three provinces.

1978 — Dr. Peter Meincke is appointed president of UPEI.

1979 — (January) Dr. Reginald Thomson of the Ontario Veterinary College begins work as planning co-ordinator for the proposed college.

1981 — Maritime provinces cannot agree on a location for the proposed college, and Thomson leaves for the Western College of Veterinary Medicine in Saskatoon.

1981 — Prince Edward Island government announces it will build the veterinary college, with or without the other provinces.

1982 — (August) Premiers John Buchanan of Nova Scotia and Jim Lee of PEI sign an agreement in principle to build the college in Charlottetown.

1983 — (February) Agriculture Minister Eugene Whelan announces that Ottawa will contribute $500,000 to begin planning and design of the proposed college.

1983 — (June) Representatives of four Atlantic provinces and federal government meet in Charlottetown to sign a cost-sharing agreement.

1983 — (July) Dr. Reg Thomson returns to Charlottetown to serve as founding dean. Construction plans begin.

1984 — (May) Sod-turning ceremony takes place.

1984 — (August) Construction begins.

1984 — (November) Council of Education of the American Veterinary Medical Association gives the college "reasonable assurance" of accreditation.

1985 — (May) Three department chairs arrive. Fourth chair (Companion Animals) appointed in May 1986.

1985 — (October) Dr. C. W. J. Eliot takes over as president of UPEI.

1986 — (September) First class at the Atlantic Veterinary College begins. Master of science program established.

1987 — (May) College dedication ceremony takes place. First graduate degree granted.

1988 — (January) Veterinary Teaching Hospital opens.

1990 — First Doctor of Veterinary Medicine degrees granted at UPEI. College gets full accreditation.

1990 — Dr. Reg Thomson resigns as dean because of ill health. Dr. Brian Hill becomes interim dean.

1991 — (October) Dr. Larry Heider becomes dean.

1994 — Animal Welfare Ynit (later Sir James Dunn Animal Welfare Centre) opens.

1995 — Dr. Elizabeth Epperly is appointed president of UPEI.

1996 — College gets full accreditation for another seven years. Class size is increased to 60.

1998 — Dr. Larry Heider is appointed acting president of UPEI. Dr. Timothy Ogilvie is appointed acting dean of AVC.

1999 — Ogilvie is appointed full-time dean. Wade MacLauchlan is appointed president of UPEI. First PhD granted.

2000 — Ottawa announces $18 million for AVC's expansion program.

2002 — Provinces agree to renew interprovincial funding agreement for fiscal years 2001–02 to 2005–06.

APPENDIX 2

Mission Statement

The Atlantic Veterinary College is committed to excellence and innovation in teaching, research and service. We believe in meeting those commitments through open communication and cooperation, in an atmosphere of honesty, accountability and mutual respect.

We exist to benefit animals, the veterinary profession and the public. Recognizing our resources and regional commitments, we aspire to provide the highest calibre of veterinary education, research and service, and to contribute to the social and economic well-being of the region.

We aim to excel at home and abroad in fields for which we are uniquely qualified.

The AVC:
- provides undergraduate, graduate, and continuing education in the field of veterinary medicine;
- conducts research to promote animal health, welfare and productivity; and to advance human and environmental health; and
- provides services and public education to improve the health and well- being of animals.

APPENDIX 3: Graduates

DVM Graduates

1990
Allyson, Kay
Barkhouse, Kim
Bos, Leanne
Bos, Paul
Charbonneau, Jane
Chong, Christina
Collard, Glen
Craig, Sylvia
Dave, Harshad K
Dingee, Carl
Duivenvoorden, Jane
Farmer, Jane
Harding, Richard
Hennessey, John
Hogan, Jane
Home, Colleen
Hughson, Susan
Kelly, Jamie
Kobalka, Joan
MacDonald, Bill
MacIssac, Paul
MacKay, Katie
MacKinnon, Cathy
Mason, Christyne
McBurney, Scott
McGregor, Leslie
Moffatt, Don
Morris, Clare
Murphy, Patty
Nicholson, Michelle
O'Brien, Tim
Paley, Dale
Perlikowski, Jerzy
Phipps, Brenda
Rideout, Jacqueline
Sanderson, David
Scott-Savage, Peter
Shaw, Eve
Smith, Jocelyn
Steele, Brent
Steen, Michael
Taylor, Dave
Teed, Carol
Thomas, Leland
Wright, Robert
Zivotofsky, Doni

1991
Adams, Skip
Barnett, Sandra
Blatt, Timothy

Boswall, James
Bulman, Suzanne
Carpenter, Wendy
Chong, Cheryl
Coldwell, Sherri
Collard, George
Collins, Kathleen
Condon, David
Cusack, Roland
deGraff, Dirk
DeHaan, Robert
Doelle, Siegmar
Foster, Janice
Gamble, Sandra
Gentile, Brian
Gilroy, Barbara
Hammond, Julia
Hartt, Randi
Heffern, Bernice
Heys, Kathleen
Hildebrand, Barbara
Kendall, Ora
Lamborghini, Amy
Levenson, Stephen
MacAulay, Edward
MacLean, Jean
Melanson, Mark
Milliken, Janis
Murnaghan, Pius
Pater, Alan
Pearce, Suzanne
Ridgeway, William
Schenkels, Wilma
Scovil, Thomas
Simms, Colleen
Sims, Heather
Stairs, Carolyn
Steele, Rebecca
Uy, Christopher
Van Donick, Helene
Van Wyk, Victoria
Williamson, David
Yole, Margaret

1992
Allain, Ann
Barnes, Janice
Bjerkelund, Susan
Bourque, Renee
Bowers, Sarah
Calder, Jill
Campbell, Claudia
Changer, Ramdath
Comeau, Yvonne

Constantine, Joanne
Crook, Janice
Cullen, Karen
Doherty, Maria
Farmer, Andrea
Gallon, Marlene
Hale, Elizabeth
Hamilton, David
Harrison, Janet
Henry, Trevor
Hitt, Nancy
Hunt, Isabel
Jones, Robert
Kelly, Jeffrey
Kerr, Robert
Lewis, Jeffrey
Mack (Irving), Anne
McCartney, Jeff
MacDonald, Jane
MacEachern, Larry
MacInnis, Eoin
MacMillan, Edward
Nieuwburg, Anja
Palmer, Colin
Peppard, Stephanie
Rodgerson, Cindy
Rutledge, Joanne
Shive, Tim
Slaunwhite, Don
Sweet, Michael
Thompson, Andrew
Vaughan, Virginia
Vessey, Gordon
Wentzell, Melissa
Wilson, Heather

1993
Alsop, Janet
Au, Judy
Babineau, Kimberley
Badcock, Jacqueline
Barrett, Bruce
Bassler, Heidi
Benoit, Donald
Brison, Monique
Canfield, Wayne
Cook, Angela
Coxhead, Brian
Crane, Ian
Dawe, Michele
DeCuba, Eric
Deluca, Adam
Dykstra, John
Gibbons, Gail

Gillis, James
Griffin, Tracey
Illingworth, Rachel
Keith, Ian
Knightly, Felicia
Kirstiansen, Margrete
Lawson, Peter
LeJenue, Jeffrey
MacKenzie, Carla
MacKenzie, Peter
MacNeill, Randy
Martin, Christopher
Mesher, Chris
More, Elliot
Morin, Vicky
Pineau-Thompson, Lisa
Rafuse, Joanne
Rice, Jody
Sabean, Eleanor
Schwartz, Anne
Stewart, Sarah
Vander Kooi, Jennifer
Vihos, Stavroula (Lula)
Wallace, Mark
Walsh, Peter
Wiliis, Michelle
Wood, Stephen
Wooldridge, John

1994
Appt, Susan
Beattie, Michael
Bercier, Yvan
Blanchard, Robbie
Bowie-Finch, Kristen
Bressler, Cindy
Byers, Leanne
Chiasson, Tim
Clarke, Lynn
Cox, Michelle
Croft, Elizabeth
Debertin, Werner
DeYoung, Cheryl
Drew, Anne
Egan, Leigh Ann
Gillis, Ann
Goodall, Jeffry
Greenlaw, Beverly
Hawkins, Leighanne
Hughes, Glenys
Jorden, Heidi
Kelly, Cheryl
Larkin, Kyra
Lavoie, Luc

LeClerc, Suzette
Legallais, Bruce
Leger, Emery
Lockhart, Kellie
Matchett, Stephanie
McDonald, Michelle
Nicol, Katherine
Pringle, Heather
Quinlan, Daniel
Raghavan, Emma
Richards, Philippa
Riffel, Scott
Rodgerson, Dwayne
Rozee, Dave
Shive, Trevin
St. Hilarie, Sophie
Theriault, Claudette
Tung, Larry
Vannan, Jan
Ward, Darryl
West, Michael
Wood, Darren

1995
Bikey, Jelal
Blasi, Valerie
Buote, Phil
Chiasson, Steven
Clark, Jodie
Creighton, Catherine
Cullen, Cheryl
DeMille, Heather
Drost, Mildred
Duizer, Glen
Elson, Chris
Eustis, John
Forsythe, Andrew
Gillis, Janice
Glencross, Juanita
Grasse, Kip
Halligan, Bridget
Hetherington, Susan
Johnson, Kobi
Jones, Sarah
Keyte, Jennifer
Legault, William
Levesque, Anie
Lomond, Susan
Low, Darren
Lowe, Janice
MacDougall, Dale
MacKay, Pamela
MacLellan, Kelly
MacLeod, Allan
McAllister, Krista
McClure, Judith
McFarlane, Rod
McPhee, Lynn

Miller, Meg
Noble, Sarah
Odian, Michael
O'Reilly, Kim
Perkins, Gillian
Piacevole, Isabella
Ravitz, Suzanne
Richter, Ruth-Anne
Roy, Jennifer
Ryan, Patricia
Silver, Gena
Sinclair, Melissa
Smith, Darrell
Stephens, Diane
Thomson-Kerr, Kim
Wiley, Valerie
Wilson, Colleen
Ziegler, Marcia

1996
Babineau, Lisa
Berton, Illina
Bourque, Andrea
Burmeister, Mayanna
Casinelle, Lisa
Cogswell, Andria
Cooper, Andrea
Corkum, Jane
Deere, Kimberley
Falkenham, Barry
Finney, Wendy
Finsten, Ariane
Frazee, Scott
Gatchell, Sara Jane
Guest, Shirleen
Hawkes, Richard
Houtsma, Greg
Kay, Sherri
Keoughan, Curry
Lake, Carmen
Lamey, Kathy
Liebenberg, Paul
MacKay, Morgan
Mack, Andrew
Maloney, Catherine
Matheson-Rakita,
  Heather
Morris, Julia
Ouellette, Marc
Parsons, Lori
Perlikowski, Anna
Petursson, Chris
Pike, Frederick
Poile, Melinda
Rogers, Laura
Ross, sheri
Sahota, Loveleen Kaur
Tasony-Ferraro, Annie

Tomplins, Adrian
Trudeau, Nadine
Wack, Oonagh
Walters, Tara Lee
Whelan, Daryl
Wilson, Roberta

1997
Adams, Susan
Allen, Catherine
Arnott, Robert
Ayers, Carolyn
Belgrave, Rodney
Campbell, James
Cizek, Susis
Cohen, Kristi
Cormier, Nicole
Dand, Jeff
Dingwell, Randy
Ferris, Mollie
Fisher, Janis
Foley, Peter
Gerras, Kimberly
Gillis, Elaine
Gilroy, Cornelia
Grant, Erin
Harris, Adrienne
Harriton, David
Hatfield, Cindy
Henderson, Janet
Hodder, Kelly
Hood, Shane
Huntsman, Lucy
Jayaram, Nandini
Kacoyanis, John
Krieger, Teresa
Langill, Alan
Lank, Trevor
Levangie, Michelle
Mailloux, Caroline
Mazzarisi, Ted
McInnis, Kelly
Michaud-Nolan, Paul
Mitchell, Kathleen
Mitchell, Krista
Mitchell, Lorri
Murphy, Marilyn
Ololick, Eva
Pfeiffer, Erika
Poirier, David
Pollock, Elizabeth
Prime, Thomas
Steele, Lesley
Suidgeest, Petronella
Voll, Rochelle
Wallace, Christa
Wanamaker, Nicole
White, Michele
Wirtanen, Erik

1998
Amos, Sarah
Bassett, Sarah
Bauman, Robert
Bragg, Natalie
Burchett, Stephen
Cleghorn, Jason
Conlon-Kelly, Ann
Corning, Michael
Cunningham, Paula
Dibblee, Suzette
Doucette, Robert
Dunnett, Margaret
Dzulinsky, Katherine
Eddy, Chicory
Fleming, Greg
Garrett, Rose Mary
Grondin, Tanya
Hollis, Carolyn
Ivany, Jennifer
Jackson, Andrew
Johnston, Stephanie
Kell, Tamara
Kuhl, Elke
MacDonald, Karieanne
MacDonald, Sara
MacEachern, Barry
MacGowan, Sheila
MacMillan, Heather
MacNeill, Catherine
McCrea-Hemphill, Lisa
McGeoghegan, Margaret
Mokler, Jody
Morris, Claude
Mosher, Heather
O'Blenis, Kelly
Osinchuk, Roger
O'Sullivan, Lynne
Phillips, Dianne
Plamondon, Suzanne
Rees, Jenny
Rowe, Michelle
Sheldon, Clare
Smith, Sherry
Smithenson, Todd
Steeves, Jennifer
Sutton, Robin
Thomas, Bronwen
Thomas, Diana
Thompson, Sarah
Vaillancourt, Jean
Wagg, Catherine
Wright, Robert

1999

Ahmad, Suzanne
Andrews, Johanna
Ayles, Heather
Bessonette, Elissa
Brown, Susan
Chen, Patty
Clancey, Noel
Crooker, Matthew
Daling, Dawn
Dooling, Nicole
Edgren, Barrett
Foote, Kim
Foreman, Krista
Fraser, Andrea
Gilbert, Jennifer
Gillot, John
Grant, Josephine
Hall, Charles
Heider, Luke
Herman, Michael
Horsman, Tricia
Johnson, Andrew
Johnson, Keith
Langham, Gregg
MacArthur, Jason
MacArthur, Susan
MacDonald, David
MacDonald, Leigh
MacDonald, Tanis
MacKenzie, Alex
Martin, Chelsea
McKenna, Shawn
McLean, Lance
McLearon, Janine
Mesher, Louise
Norman, Brian
Owens-Evans, Lisa
Perkins-Masters, Teri-
  Lynn
Perry, Tanya
Ross, Ian
Roxborough, Heather
Rudich, Stephanie
Rundle, Tracie
Sartor, Laura Lee
Saxon, Meg
Schneider, Amy
Simons, Helen
Sullivan, Bradley
Taylor, Stephen
  (Posthumous
  Honorary)
Verschoor, Marc
Wheeler, Michael

2000

Ansems, Marian
Bailey, Trina
Beazley, Shelley
Bourque, Troy
Boutilier, Pam
Brennan, Sebastian
Davidson, Ainsley
Dewitt, Shane
Dhama, Vinay
Dickinson, Ryan
Doyle, Aimie
Fernandez, Nicole
Forgeron, Celeste
Fraser, Audrey
Furey, Karla
Galbraith, Craig
Hart, Maarten
Hotham, James
Julien, David
King, Bradley
Lahey, Julie
Landry, Tracy
Lewis, Shannon
Logan, Miranda
MacDonald, Christopher
MacDonald, Scott
MacDonald, Valerie
MacKay, Chris
MacKay, Lisa
MacLean, Heidi
MacWilliams, Christine
Maimon, Elizabeth
Marche, Candace
Marquis, Patricia
Marr, Tiffany
Marshall, James
Matchett, Kelly
McEvoy, Madeline
Miller, Barry
Oakley, Michelle
Sampathkumaran,
  Raghavan
Schaefer, Nicole
Scott, Jessica
Shaw, Carolyn
Silverstone, Andrew
Strong, Shawna
Sutherland, Lynn
Szyngiel, Betty
Tokiwa, Michael
Turner, Kari
Whalen, Lisa
Williams, Laurel

2001

Allen, Amy
Allen, Roselyn
Anthony, Michele
Blake, Justin
Burke, Deborah
Caines, Jane
Carboni, Deborah
Clarke, Trisha
Davis, Timothy
Dunn, Danielle
Eifler, Danielle
Fishkin, Randi
Galvin, Marie-claire
Godbey, Tamara
Hackett, Karen
Hanelt, Lisa
Hendsbee, Jill
Higgins, Janet
Higgins, Tanya-Marie
Hochman, Daniel
Hockman, Cynthia
Jamieson, Lara
Kallen, Shana
Katz, Joann
Kent, Any
Keppie, Nathan
Klinck, Mary
Lagace, Daniele
Lariviere, Carolyn
Larkin, David
Leblanc, Mireille
Leger, Clare
Lewis, Stephen
Ling, Kathy
Little, Natasha
MacDonald, Rosalyn
MacLeod, Andrew
MacMillan, Kathleen
Major, Lisa
Marshall, Beth
Melvin, Ellen Rae
Meredith, Taralyn
Monteith, Shannon
Murray, Cari
Newcomb, William
Nicki, Alexander
O'Brien, Nicole
Piercey, Suzanne
Pottie, Gregory
Proud, Sarah
Robblee, Frank
Rondeau, Fabienne
Runnalls, Angela
Ryan, Shanna
Schwartz, Pamela
Spears, Jonathan
Surette, Justin

Thompson, Gawen
Young, Jennifer
Zafiris, Neove

2002

Becker, Melissa
Beebe, Dale
Bihr, Tanja
Bornstein, Danielle
Castaneda, Leia
Chapman, Tara
Choker, Dru
Crane, Bronwyn
Davies, Sandra
Fawcett, Loridawn
Galbraith, Laura
Giusti, Michael
Golding, Christine
Harvey, Jessica
Inness, Kyla
Jakubiak, Martin
Keizer, Karen
King, Ryan
Kirk, Simon
Koch, Kristina
Kohout, Douglas
La Stella, John
Lavers, Carrie
Lawrie, Ian
Lees, Robert
Legge, Andrea
Leonard, Erin
MacKenzie, Terri
Magerkorth, Johnathan
Maguire, Fintan
Mahon, Melanie
Martin, Dawn
Matchet-Waterhouse,
  Amy
Matheson, David
Matsumoto, Heather
Matthews, Tracy
Miller, Justin
Muirhead, Tammy
Mulley, Kelli
Murrya, Heather
Nicholson, Peter
Oulton, Robert
Peacock-Phillips,
  Barbara
Plosky, Helen
Powell, Leslie
Rideout, Alison
Riggs, Cherie
Rutherford, Erin
Saulais, Danielle
Smyrni, Keri
Stojanovic, Vladimir

ten Broeke, Tanya
Thomas, Stacey
Tripi, Christopher
Unger, Teri
Vaughan, Andrew
Walker, Ingrid
Wilcox, Shayne
Williams, Meghan

2003
Adams, Samantha
Baker, Heather
Bennett, Jennifer
Bergmann, David
Blundell, Tracey
Boyer, Lauren
Buote, Melanie
Carlisle, Mary Ellen
Copp, Catherine
Currie, Ailsa
D'Amato, Dan
Emdin, Pamela
Emo, Toby
Evers, Keri
Fawcett, Brent
Francheville, Carla
Galanthay, Susan
Geariety, Stephanie
Gehrig, Michiko
Goetz, Jennifer
Giffin, Pauline
Guck, Ilse
Hall, Meagan
Hare, Alex
Hartnett, Leighann
Henderson, Tya
Hicks, Melanie
Johnson, Heather
Jones, Patti
Kadri, Isaam
Keirstead, Natalie
Lear, James
Lewis, Kristine
Lister, Stephanie
Loeman, Herb
MacIntyre, Juanita
McKay, Jennifer
Marryatt, Paige
McCullum, Darcie
Meister, Babette
Mroziak, Melissa
Munn, Sarah
Murphy, Fidel
Murrell-Liland, Blakeley
Nelson, Liane
Niblett, Kim
O'Neil, Betsy
Olafson, Kristin

Pedersen, Victoria
Prescesky, Joy
Purje, Anneli
Rawlins, Danielle
Russell, Deanna
Russo, Mark
MacDonald, Kimberley
Sheehan, Karen
Strassner, Mike
Volkers-Hillman, Mary
Welland, Lisa

2004
Aucoin, Melissa
Avery, Stephanie
Black, Emily
Blois, Shauna
Bottoms, Bartholomew
Boulter, Andrew
Brown, Sara
Butler, Jarrod
Churchill, Vanessa
Corfield, Kim
Cornell, Alison
Cox, Brandie
Emmans, Erin
Emmett, Cheala
Gaylord, Maureen
Grant, Hanna
Graves, Brian
Hamilton, Rhonda
Haney, Christine
Hayes, Felicia
Hindson Jenkin, Lisa
Jones, Barbara
Jones, Katherine
Kavanagh, Victoria
Keenan, Samantha
Kline, Leanne
Lachance, Isabelle
Laite, Cheryl
Lam, Heather
Landry, Marie
Landry, Tania
Lawson, Trevor
Mallov, David
Manfredi, Jane
Martinson, Shannon
Martyn, Angela
Moffett, Patricia
Orr, Jeremy
Osmond, Maureen
Parkes, Jennifer
Perry, Jeanette
Piscitelli, Christopher
Pringle, Emily
Puddester, Krista
Ruigrok, Wendy

Saunders, Janet
Schrage, Amanda
Sin, Catherine
Spencer, Penny
Steele, Karyn
Stewart, Kendra
Swim, Amanda
Swinamer, Sonya
Vanderstichel, Raphael
Williams Mason, Alana
Williamson, Leigh
Yazer, Joseph

MSc Graduates

1989
Bernardo, Theresa
Hanna, Paul

1990
Bildfell, Robert
Davidson, Jeffrey
Drake, Wendy
Gallant, Richard
Grimmelt, Bryan
Liu, Jiang
Mitton, Gregory
Power, Carl
VanTil, Linda
Wadowska, Dorota

1991
Blanchard, Robert
Burton, Shelley
Cawthorn, Elisabeth
Donovan, Bernadette
Sidhu, Nirmal

1992
Chohan, Amrit
DeVries, Wilma
Dohoo, Susan
Hurnik, Daniel
James, Cheryl
Owens, Kelli
Powell, Mark
Singh, Chandrapal
Vokaty, Sandra

1993
Anderson, Stewart
Brown, Jennifer
Bruneau, Nancy
Chibeu, Malanga
Corkum, Mary Jane
Drastini, Yatri
Dybing, Jody
Giles, Janice

Hammell, Lawrence
Hitt, Mark
Hood, Rosemary
Klein, Thomas
Pye-MacSwain, Joy
Zhou, Chengfeng

1994
Bellamy, Craig
Crossley, Jennifer
Donovan, Arthur
Gillis, Janice
Kibenge, Molly
Murphy, Shane
Murray, Harry
Ndiritu, Wangeci M.
Newbound, Garret
Nganda, Gatei wa.
Omukuba, John

1995
Huntsman, Lucy
Keefe, Gregory
Xia, Jane

1996
Dhama, Vinay
M'Aburi, Karega
McBurney, Kim
Mullins, Julia
Nagarajan, Malliga
Omar, Semir
Peng, Jian
Plourde, Susan
Rostant, Elizabeth
Underhay, John

1997
Ahmed, Hussein
Almendras, Felipe
Connell, Barry
MacIsaac, Paul
Miller, Tanya
Sanchez, Genaro

1998
Baglole, Carolyn
Ficele, Giovanni
LeBlanc, Deborra
MacCallum, Jillian
Munroe, Fonda
Schurman, Robert
Trevors, Crystal

1999
Abouzeed, Yousef
Beaman, Holly
Bihr, Tanja
Carjaval, Veronica
Lavallee, Jean
Morrison, Jennifer
Pustowka, Cory
Salinas, Everardo
Wartman, Cheryl
Zhou, Jianwei

2000
Campbell, Pat
Cook, Todd
Foote, Kim
Gonzalez, Pablo
Graham, Kristin
Guy, Norma
Knight, Joy
MacMillan, Kathleen
Morrison, Sandie
Ramaswamy,
    Chidambaram

2001
Forzán, Maria
Gardiner, Yvette
Joseph, Tomy
Ramsay, Jennifer

2002
Arunvipas, Pipat
Bowers, Joanne
Brake, John
Courtland, Hayden-
    William
Fast, Mark
Leger, Emery
McQuaid, Tim
Nevárez, Alicia
Nødtvedt, Ane

2003
MacLean, Linda
Marshall, Rebecca
Murphy, Gailene
Nie, Jingbai
Rigley, Sarah
Udayamputhoor, Roy
Vijarnsorn, Monchanok

2004
Christie, Julie
Craig, Sylvia
Parkyn, Geoff
Waite, Linda
Wojciechowska, Janina

MVSc Graduates

2004
Haruna, Julius

PhD Graduates

1999
Hovingh, Ernest

2000
Aburto, Enrique
Kibenge, Molly
Martinez-Burnes, Julio
Nagarajan, Malliga

2001
Sánchez-Martinez,
    Genaro

2002
Gaskill, Cindy

2003
Battison, Andrea
Bedard, Karen
Doucette, Tracy
Nanton, Dominic
Rodriquez, Juan Carlos
Rodriquez, Luis

2004
McClure, Carol
Sanchez, Javier

APPENDIX 4

The R. G. Thomson Academic Achievement Medal

This award is given to the graduating student in veterinary medicine at the Atlantic Veterinary College with the highest cumulative grade average for the four-year program.

Recipients

1990	Sylvia Craig
1991	Margaret Yole
1992	Janice Barnes
1993	Eleanor Sabean-Walsh
1994	E. Anne Drew
1995	Cheryl Cullen
1996	Oonagh Wack
1997	Peter Foley
1998	Lynne O'Sullivan
1999	Elissa Bessonette
2000	Shelley Beazley
2001	Sarah Proud
2002	Andrew Vaughan
2003	Natalie Keirstead
2004	Shannon Martinson

APPENDIX 5

Chairs' Terms of Office

Pathology & Microbiology

James Bellamy	1985–88
Ray Long	1988–94
Frederick Markham	1994–2000
Basil O. Ikede	2000–present

Companion Animals

Brian Hill	1986–98
David Seeler	1989–90 (Acting)
Dana Allen	1998–99
James Miller	1999–2004
Darcy Shaw	2004–present

Health Management

Robert Curtis	1985–90
Tim Ogilvie	1990–98
Ian Dohoo	1993–94 (Acting)
Elizabeth Spangler	1998–2004
Jeff Wichtel	2004-present

Anatomy & Physiology/Biomedical Sciences

James Amend	1985–93
Andrew Tasker	1993–99
Bill Ireland	1996 (Acting)
Luis Bate	1998–99 (Acting)
Luis Bate	1999–present

Appendix 6: AVC Tenure-Stream faculty

NAME	DATE OF INITIAL APPOINTMENT	DATE OF RESIGNATION OR RETIREMENT
Biomedical Sciences		
Amend, Jamie	May 1, 1985	August 28, 1995 (resigned)
Bate, Luis	June 1, 1986	
Burka, John	July 1, 1985	
Chan, Catherine	July 1, 1988	
Cribb, Alastair	July 1, 1996	
Dawson, Susan	July 1, 1996	
Dohoo, Susan	October 1, 1992	
Gaskill, Cynthia	September 1, 2001	
Hewson, Caroline	October 30, 2000	
Ireland, William	June 1, 1987	
Kamunde, Collins	January 1, 2004	
Nijjar, Mohinder (Peter)	June 1, 1987	June 30, 2001 (retired)
Novotny, Mark	October 1, 1987	September 18, 1995 (resigned)
Saleh, Tarek	August 1, 1996	
Sims, David	July 1, 1986	
Singh, Amreek	July 1, 1985	June 30, 2000 (retired)
Stokoe, William	September 1, 1985	September 30, 1995 (retired)
Tasker, Andrew	August 1, 1987	
Walshaw, Sally	September 9, 2002	
Wright, Glenda	July 1, 1988	
Companion Animals		
Allen, Dana	September 14, 1998	June 25, 1999 (resigned)
Bailey, Trina	January 15, 2006	
Basher, Anthony	July 15, 1987	June 30, 1996 (resigned)
Cullen, Cheryl	August 15, 2000	
Curtis, Michael	August 1, 1991	May 29, 1996 (resigned)
Gelens, Hans	July 1, 1998	
Gibson, Karen	July 1, 1989	November 2, 2001 (resigned)
Hill, Brian	May 1, 1986	January 15, 2000 (deceased)
Hitt, Mark	August 1, 1986	July 15, 1992 (resigned)
Hogan, Patricia	July 1, 1987	January 20, 1996 (resigned)
Ihle, Sherri	November 1, 1990	

Kaderly, Robert	July 1, 1991	September 1, 1995 (resigned)
Lamont, Leigh	January 7, 2002	
Lane, India	October 15, 1992	August 31, 1997 (resigned)
Lemke, Kip	January 18, 1993	
Marohn, Mark	July 1, 1989	December 6, 1992 (resigned)
Mattoon, John	July 1, 1997	June 1, 1999 (resigned)
Miller, James	July 1, 1989	
Moak, Peter	November 3, 2002	
Moens, Noel	July 1, 1997	July 31, 2000 (resigned)
Pack, LeeAnn	August 15, 2001	
Rose, Patricia	May 4, 1998	
Runyon, Caroline	May 1, 1986	
Seeler, David	March 1, 1987	
Shaw, Darcy	August 1, 1987	
Solano, Mauricio	September 1, 1994	June 1, 1997 (resigned)
Walshaw, Richard	September 9, 2002	

Health Management

Barkema, Herman	August 1, 2001	
Brimacombe, Michael	July 1, 1996	September 1, 2000 (resigned)
Curtis, Robert	May 1, 1985	September 11, 1996 (retired)
Davidson, Jeffrey	September 1, 1988	
Desjardins, Marc	November 1, 1989	June 30, 1997 (resigned)
Dohoo, Ian	July 1, 1985	
Donald, Alan	July 1, 1986	December 31, 1995 (resigned)
Duckett, Wendy	November 3, 1997	
Ehrlich, Paula	September 1, 1995	February 28, 1999 (resigned)
Hammell, Larry	September 15, 1992	
Heider, Lawrence*	October 1, 1991	June 30, 1999
Huhn, John	September 1, 1987	July 31, 1994 (resigned)
Hurnik, Daniel	April 1, 1989	
Jansen, Maura	May 1, 1990	December 13, 1991 (resigned)
Keefe, Gregory	February 1, 1996	
Lofstedt, Jeanne	November 1, 1987	
Lofstedt, Robert	November 1, 1987	
McClure, J. Trenton	September 7, 1998	
McDuffee, Laurie	September 1, 1998	
McNiven, Mary	July 1, 1986	

* Acting UPEI President from August 15/98 to June 30/99

Ogilvie, Timothy	July 1, 1985	
Ortenburger, Arthur	September 1, 1987	
Pringle, John	July 1, 1989	April 30, 1997 (resigned)
Richardson, Gavin	September 1, 1987	
Riley, Christopher	April 1, 2001	
Ruegg, Pamela	July 1, 1989	June 30, 1992 (resigned)
Scharko, Patricia	July 1, 1989	October 20, 1995 (resigned)
Slana, Michael	December 7, 1992	January 1, 1996 (resigned)
Spangler, Elizabeth	January 1, 1987	
Stryhn, Henrik	September 1, 2001	
VanLeeuwen, John	August 1, 1999	
Wichtel, Jeffrey	February 2, 1998	

Pathology and Microbiology

Bellamy, James	May 1, 1985	
Burton, Shelley	October 15, 1993	
Casebolt, Donald	September 1, 1993	May 31, 1998 (resigned)
Cawthorn, Richard	July 1, 1985	
Cepica, Arnost	July 1, 1985	
Conboy, Gary	August 1, 1991	
Daoust, Pierre-Yves	October 1, 1987	
Florence, Darrel	March 1, 1999	February 29, 2000
Fuentealba, Carmencita	September 1, 1992	August 31, 2001 (resigned)
Hanna, Paul	July 1, 2002	
Hariharan, Harihar	January 1, 1988	
Hobert, Eric	July 1, 1989	September 14, 1990 (resigned)
Honor, David	August 1, 1987	November 13, 1992 (resigned)
Horney, Barbara	August 1, 1992	
Ikede, Basil	September 1, 1989	
Johnson, Gerald	September 1, 1986	
Kibenge, Frederick	June 1, 1989	
Lewis, Jeffrey	May 15, 1998	
Long, Ray	July 1, 1986	June 30, 1995 (retired)
Lopez, Alfonso	August 1, 1988	
Markham, Frederick	July 1, 1986	
Miller, Lisa	July 1, 1989	
Speare, David	September 1, 1990	
Thomson, Reg	July 1, 1983	January 21, 1991

Appendix 7: Examples of AVC Faculty Receiving National or International Recognition

Faculty Member	Academic Department	Recognition
Dr. Franck Berthe	Pathology & Microbiology	Canada Research Chair (2004–2009)
Dr. John Burka	Biomedical Sciences	Gordin Kaplan Award for Science Awareness, Canadian Federation of Biological Sciences (1997)
Dr. Alastair Cribb	Biomedical Sciences	Canada Research Chair (2001–2006)
Dr. Bob Curtis	Health Management	CVMA Lifetime Membership (2003)
Dr. Sue Dawson	Biomedical Sciences	Student AVMA Teaching Excellence Award in Basic Sciences (2000), (chosen from all US and Canadian veterinary colleges)
Dr. Dan Hurnik	Health Management	Schering-Plough Animal Health Award (1999), CVMA
Dr. Karen Gibson	Companion Animals	CVMA Humane Award (2000)
Dr. Ian Dohoo	Health Management	Schering-Plough Animal Health Award (1995), CVMA
Dr. Sherri Ihle	Companion Animals	Daniels Award for Excellence in the Advancement of Knowledge Concerning Small Animal Endocrinology (1992)
Dr. Greg Keefe	Health Management	Schering-Plough Animal Health Award (2002), CVMA
Dr. Fred Kibenge	Pathology & Microbiology	NSERC's 25th Anniversary Honour (2004)
Dr. Caroline Runyon	Companion Animals	Northeast Regional Service Award, AAHA (1993)
Dr. Sally Walshaw	Biomedical Sciences	Bustad Companion Animal Veterinarian of the Year Award (1994), AVMA
		Carl J. Norden Distinguished Teacher Award, Pfizer Canada Inc.
Dr. Lisa Miller	Pathology & Microbiology	1993
Dr. William Ireland	Biomedical Sciences	1994
Dr. Carmen Fuentealba	Pathology & Microbiology	1995
Dr. Shelley Burton	Pathology & Microbiology	1996
Dr. Amreek Singh	Biomedical Sciences	1997
Dr. Alfonso Lopez	Pathology & Microbiology	1998
Dr. Jim Miller	Companion Animals	1999
Dr. Kip Lemke	Companion Animals	2000
Dr. Ian Dohoo	Health Management	2001
Dr. David Speare	Pathology & Microbiology	2002
Dr. Art Ortenburger	Health Management	2003
Dr. Greg Keefe	Health Management	2004

APPENDIX 8

Examples of Research Projects or Programs at the Atlantic Veterinary College

Dr. Luis Bate
Use of vocalizations to enhance productive processes in animals
Broadcasting the respective species specific feeding calls to newly hatched birds as soon as they are placed in a barn. Broadcasting pre-recorded feeding calls at specific intervals will attract birds (poults and chicks) to initiate feeding. Growth is enhanced by 12–25% over the first three weeks of life. A poultry feeder capable of delivering prerecording vocalizations has been patented.

Microwave Technology
1) A rewarming technique for hypothermic piglets using 915 MHz microwaves, which can raise core temperature at rates of 1C/min. This technology has the potential to be applied to humans.
2) Supplementary heating based on microwaves maintains animals' comfort, as well as reduces agonistic behaviour without heating the air of the room, which is then ventilated out. This technology has the potential of impacting the energy requirement of future housing systems.

Dr. John Burka
The Interaction Between Sea Lice and Salmonid Fishes: Development of Alternative Methods of Control
Collaborative efforts among AVC scientists, the NRC Institute for Marine Biosciences, and the Aquaculture Industry to examine physiological and immunological interactions between sea lice and salmon so that sea lice can be controlled more effectively. J. F. Burka was Principal Investigator for a five-year grant from the NRC/NSERC Research Partnership Program. This project resulted in 18 papers, 7 reports, and 2 book chapters.

Development of strategies to identify, monitor, and control resistance of sea lice to therapeutants used in salmonid aquaculture
These are collaborative efforts among scientists in Canada and Eu-

rope and the Aquaculture Industry to increase our understanding of the mechanism of action of drugs being used for controlling sea lice, carried out in conjunction with epidemiological studies documenting efficacy and treatment failures.

Dr. Rick Cawthorn

Dr. Rick Cawthorn is Professor of Parasitology in the department of Pathology and Microbiology, and founding Director and Senior Scientist of the **AVC Lobster Science Centre**™. After his arrival in Charlottetown in May 1985 as one of the first faculty members of AVC, his research interests shifted to parasitic protozoa of fish, including crustaceans, bivalves, and finfish. Rick has led the research program on crustacean health, beginning in spring 1994.

Dr. Alastair Cribb

Dr. Alastair Cribb's research program in drug and chemical safety seeks to understand the molecular and genetic basis of species and individual susceptibility to adverse drug reactions and breast cancer risk. His work in animals and humans has encompassed sulfonamide hypersensitivity reactions, phenobarbital toxicity, the role of the endoplasmic reticulum in drug and chemical toxicity, and the role of estrogen in breast cancer risk and in protecting neurons following stroke.

Dr. Wendy Duckett

Along with Pat Rose and Ian Dohoo, Dr. Wendy Duckett has worked on the early markers of equine joint disease. She has also investigated the circadian rhythm of equine thyroid hormones. Dr. Duckett is currently working on a project in England with phage researchers on using bactertiophage against E. coli and Salmonella bacteria.

Dr. Cynthia Gaskill

Investigation of Adverse Effects of Anticonvulsant Drugs in Dogs

Phenobarbital and potassium bromide are the most common anticonvulsant drugs used to treat epilepsy in dogs. Both drugs are associated with a number of adverse effects that can necessitate discontinuation of the drugs or even result in death. We are investigating risk factors and mechanisms by which these adverse effects occur in order to better prevent and treat them.

Investigation of Lung Disease Associated with Potassium Bromide Therapy in Cats

Potassium bromide is occasionally used to treat seizure disorders in cats. However, development of respiratory distress and even death have been reported in some cats receiving the drug. We are investigating the mechanism by which this adverse effect occurs in order to better understand the problem and to potentially develop a model for respiratory diseases involving ion transport disorders.

Dr. Cornelia Gilroy, Dr. Shelley Burton, Dr. Barbara Horney, Allan MacKenzie

Validation of Ionized Magnesium Measurement in Feline Sera Using the Nova CRT8

Evaluation of ionized magnesium concentrations may provide optimal patient care. Cats experience diseases that could alter these concentrations, such as renal disease and hyperthyroidism. This project evaluated the stability, precision, linearity, analytical sensitivity, and reference interval for an analyzer and assay of ionized magnesium in cats. It proved to be reliable, but utilization requires special blood sample handling.

Dr. Norma Guy

Canine Household Aggression

A survey of dog owners in the Maritime provinces gathered information on aggressive behaviour toward family members, the most common form of aggression in dogs. This research was groundbreaking in its design, leading to an ability to identify significant risk factors. It confirmed previous reports of a possible association between aggression and spaying in female dogs, and refuted the belief that the majority of canine aggression is related to problems with dominant behaviour.

AVC Humane Dog Training Program

This service project has provided veterinary students to work as dog trainers and adoption counsellors at the PEI Humane Society for the past four years. The students working in the shelter also provide much-needed social enrichment for the shelter dogs. It is an excellent opportunity for future veterinarians to learn more about the issues faced by all shelters, while at the same time improving the welfare of homeless dogs.

Dr. Caroline Hewson

Assessment of the Quality of Life of Pet Dogs

We have developed a preliminary instrument for assessing non-physical aspects of pet dogs' quality of life. Such assessment is important because it could help dog owners and their veterinarians make informed decisions concerning dogs' well-being (e.g., cancer treatment; moving house). Quality of life research is new in veterinary medicine and our study has revealed many important questions about how best to assess pets' quality of life.

Canadian Veterinarians' Use of Painkillers in Dogs and Cats After Surgery

This project is a repeat of an innovative one conducted by Susan and Ian Dohoo of the AVC in 1994. They conducted a national, representative survey of veterinarians, using a mailed questionnaire to find out how veterinarians managed pain in their surgical patients. At the time of writing, statistical analysis is nearing completion. The results will help to inform veterinary teaching and continuing education.

Dr. Basil Ikede

In 2000, Dr. Basil Ikede, professor of pathology, proposed a **non-thesis MVSc program and a Postgraduate Diploma program** with the overall goal of providing advanced professional training to veterinarians. He chaired a committee that saw through the approval of both programs by 2001. A year later, four students were admitted for the MVSc in pathology. UPEI's first MVSc degree was awarded in 2004 and the program has been adopted in the department of Health Management.

Dr. Alfonso Lopez
Meconium Aspiration Syndrome is an important condition in pediatric medicine and veterinary neonatology. Our laboratory developed an experimental model, and we are currently studying the patholgenesis and effects of this syndrome on lung structure and function.

The Animal Health, Aquaculture and Diagnostic Services was the first major project funded by the **Canadian International Development Agency** at UPEI. This six-year project with a total budget of $1.4 million was implemented to improve the quality of education and services at the University of Tamaulipas, Mexico. Over 18 Mexican faculty were trained at AVC, and 14 AVC faculty members provided hands-on training courses at the University of Tamaulipas in Mexico.

Dr. Mary McNiven
Nutritional research has been carried out by Dr. Mary McNiven. Work has included **the evaluation and improvement of the nutritional quality of locally produced feedstuffs for dairy and beef cattle, swine, poultry, and aquaculture species**. In addition, she is investigating ways to improve the healthfulness of animal products by altering the fat composition, and by using the bioactive compounds in blueberries to improve shelf-life of fish products.

Dr. Gavin Richardson, Dr. Mary McNiven
Dr. Gavin Richardson and Dr. Mary McNiven have studied **sperm metabolism and viability during long- and short-term storage in horses, pigs, and fish**. Work has included utilization of ATP in sperm over time, membrane integrity and capacitation status, as well as the use of antioxidants in storage media. The interaction between nutrition and reproduction was studied with rainbow trout by manipulating the fatty acid composition of sperm membranes to improve viability after cryopreservation.

Dr. Tarek Saleh

Since joining UPEI in 1996, Dr. Saleh has investigated areas including **how the brain controls blood pressure and heart rate and the cardiac consequences of cardiovascular diseases**. He has received over a million dollars from external funding sources including the Heart & Stroke Foundation of PEI (PI), the Canadian Institutes for Health Research (PI), Alzheimer Society of Canada (collaborator), and the Canadian Breast Cancer Research Initiative (collaborator). Dr. Saleh has also published over 35 peer-reviewed journal articles.

The Role of Neurochemicals and Neurosteroids

In Dr. Tarek Saleh's most recent work, he has been exploring the role that neurochemicals and neurosteroids play in cardiovascular regulatory regions in the brain. The overall objective of this research program is to elucidate the neural mechanisms that modulate cardiac function before, during, and after the development of cardiovascular disease in males and females.

Dr. David Sims

David Sims joined the AVC faculty in 1986. His research has been primarily based on ultrastructure of the microvasculature in inflammation, with departures during the domoic acid incident of the late 1980s and to collaborate in fish research in the 1990s. Heart and Stroke Foundation of Canada and National Institutes of Health (USA) funding were obtained to investigate **the roles of pericytes in reducing inflammatory reactions**.

Drs. Dave Speare, Fred Markham, & Jeff Lewis

Project Loma, led by a multidisciplinary research team, has received eight years of funding (totaling more than $1 million) through the NSERC Strategic Grant Program. The program has provided research training for graduate students and post-doctoral researchers here at the AVC and other collaborative research centres in British Columbia; research findings have advanced our understanding of the life cycle of *Loma salmonae* and have provided treatment and vaccination options for the salmon farming industry.

Dr. Andy Tasker
Characterization of the in vivo pharmacology of domoic acid
Since 1987, domoic acid (DOM) has become a biotoxin of world-wide concern. It has also emerged as a valuable pharmacological tool for neurobiological research. My lab is recognized internationally for our investigations of the toxicological and pharmacological profile of DOM in rodents. This work, continually funded by federal granting councils (NSERC and CIHR) since 1989, has resulted in over 60 publications and programs for 15 trainees.

Kainate receptors in brain development
Kainate receptors (KAR) are critically involved in normal brain function and neurological disease. With over $0.7 million in support from CIHR and NSERC, his lab has published many pioneering studies on KAR involvement in brain development in rats. This has resulted in numerous papers and symposia as well as two patent applications. It is also one of three core research programs in the new Atlantic Centre for Comparative Biomedical Research.

Dr. Maureen Wichtel (UPEI), Dr. J. Wichtel (UPEI), Dr. G. Guamme (UBC), Dr. Larry Hale (UPEI), and Stephanie Power (Honours Student, UPEI)
Development of a Novel Assay for Detection of Hypomagnesemia in Horses
This assay appears to reflect intracellular magnesium concentration that has been developed (the MRG1 assay). The objective is to establish whether MRG1 expression may be used as an indicator of cellular balance in horses, as has been found previously in mice and humans. The development of this test for domestic species will lead to a better understanding of how diet and disease affect magnesium status.

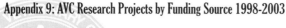

Appendix 9: AVC Research Projects by Funding Source 1998-2003

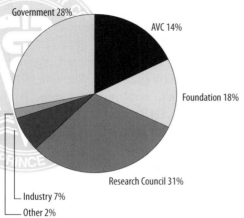

AVC faculty conduct a varied program of research, including basic biomedical research and animal health research in fish, lobster, swine, dairy, beef cattle and other animals. Revenues from on-going research amounted to over $4 million in 2002-2003. An analysis of research funds for the period 1998-2003 indicates that 31% came from the major research councils (such as NSERC and CIHR), 28% came from other government sources, 18% came from various foundations, 14% from AVC, and 7% came from industry.

Appendix 10: AVC Research Funding by Subject 1998-2003

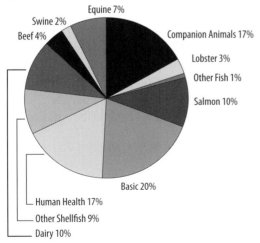

APPENDIX 11

Atlantic Veterinarians by Type of Practice, March 2004					
	NB	PE	NS	NB	Atlantic
Small Animal	31	48	123	101	303
Mixed Animal	6	47	95	20	168
Large Animal	4	26	12	25	67
Industry	0	1	1	1	3
Government	9	5	14	17	45
Teaching	0	27	2	1	30
Research	0	3	1	0	4
Other	2	15	14	7	38
Retired	4	5	6	15	30
Total	56	177	268	187	688

(Source: Canadian Veterinary Medical Association)

APPENDIX 12

Potential Impact of AVC on Selected Atlantic Industries, 1999					
	NFL $million	PEI $million	NS $million	NB $million	Atlantic $million
Livestock (farm incomes)	62.7	104.0	239.3	179.0	585.0
Aquaculture (output value)	17.7	22.7	27.9	157.7	226.0
Lobster (landings)	19.0	82.6	236.3	75.2	413.1
Total income or output	**99.5**	**209.3**	**503.5**	**411.9**	**1224.1**
AVC provincial grant	0.5	5.2	3.8	3.1	12.6
Required impact on industry income to "justify" AVC provincial funding	0.5%	2.5%	0.8%	0.8%	1.0%
Note: Value of lobster landings are for 1998.					

(Source: APEC calculations using data from AVC and Statistics Canada)

Appendix 13: AVC Provincial Grants as % of Total Annual Revenues
1995-1996 to 2003-04

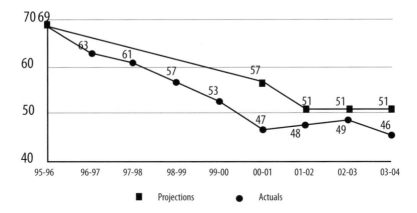

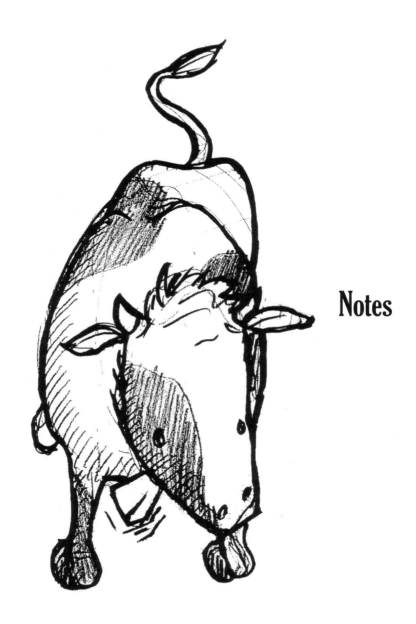

Notes

Chapter 1

1 F. Eugene Gattinger, *A Century of Challenge, A History of the Ontario Veterinary College*, University of Toronto Press, 1962.
2 Report of the Prince Edward Island Veterinary Medical Society, submitted to the PEI Advisory Reconstruction Committee (1945).
3 J. A. Archibald, T. J. Hulland, H. N. Vance, *Veterinary Manpower in Canada, Report of the Committee of Enquiry to the Council of CVMA*, July 1977.
4 T. L. Jones and W. A. Moynihan, *National Veterinary Manpower Survey*, Ottawa, 1971.
5 *Veterinary Manpower in Canada.*

Chapter 2

1 *Veterinary Manpower in Canada.*
2 G. Edward MacDonald, *If You're Stronghearted, Prince Edward Island in the Twentieth Century* (Charlottetown: Prince Edward Island Museum and Heritage Foundation, 2000), p. 259.
3 Edward MacDonald, *The History of St. Dunstan's University, 1855–1956* (Charlottetown: Board of Governors of St. Dunstan's University and the Prince Edward Island Museum and Heritage Foundation, 1989), p. 263.
4 Because of the passage of time and absence of documentation, opinions differ on exactly what transpired at this point in the AVC saga. What follows is my best attempt to untangle the course of events.
5 Ron Baker, letter to *The Guardian*, Nov. 8, 1974.
6 D. G.Howell, Report of a Study of the Establishment of a School of Veterinary Medicine in the Atlantic Region, Aug. 31, 1975.
7 Ibid.
8 Now called the Maritime College of Forest Technology.
9 Howell, *op cit.*
10 Ibid.
11 Ibid.
12 Letter from Dennis Howell to Premier A.B. Campbell, Nov. 16, 1976.
13 Letter from Eugene Whelan to Premier A. B. Campbell, Feb. 21, 1977.

Chapter 3

1 MPHEC annual reports.
2 Prince Edward Island contributed more capital and operating dollars than Nova Scotia, but not the majority of total funding.
3 Rolland Hayman, letter to Eugene Whelan, September 27, 1977.
4 "Can Nova Scotia really afford a new veterinary college?" Halifax *Chronicle-Herald*, November 8, 1979.
5 Dr. D. L. T. Smith, letter to Tom McMillan, MP, September 9, 1980.
6 J. A. Archibald, T. J. Hulland, H. N. Vance, *Veterinary Manpower in Canada, Report of the Committee of Enquiry to the Council of CVMA*, July 1977
7 Dennis Howell, letter to Eugene Whelan, June 4, 1981.
8 Clayton MacKay, letter to Eugene Whelan, Aug. 1, 1980.
9 Minutes, meeting of Council of Maritime Premiers, June 16, 1980.
10 Minutes, Council of Maritime Premiers meeting, September 22–23, 1980.
11 R. G. Thomson, memorandum to David Weale, October 27, 1980.
12 Angus MacLean, letter to John Buchanan, December 2, 1980.
13 R. G. Thomson, letter to Dr.W.M. Adams, January 16, 1981
14 Statement of J. Angus MacLean regarding the proposed Atlantic Veterinary College, Angus MacLean papers in PAPEI.

[15] J. Angus MacLean, letter to Eugene Whelan, April 1, 1981.
    Also, Peter Meincke, letter to M. A. (Bud) Olson, July 29, 1981.
[16] Dennis Howell, letter to Eugene Whelan, June 4, 1981.
[17] Peter Meincke to Eugene Whelan, *op cit.*

Chapter 4

[1] Interview with Jim Lee, July 26, 2002.
[2] Eugene Whelan telex to Prowse Chappell, Feb. 24, 1983; *House of Commons Debates*, April 1, 1982.
[3] Richard Duffy, Dean Shaw, Arthur Howard, Keith Mullins, letter to Eugene Whelan, May 19, 1982.
[4] Peter Meincke, letter to John Pickard, May 14, 1982.
[5] Jim Lee, letter to John Pickard, May 12, 1982
[6] *House of Commons Debates*, April 1, 1982.
[7] *House of Commons Debates*, May 18, 1982.
[8] Tom McMillan, speaking at the dedication of AVC, May 9, 1987.
[9] Henry Phillips, memo to Jim Lee, January 21, 1982.
[10] Papers of Jim Lee, Public Archives of PEI.
[11] Eugene Whelan, letter to Bennett Carr, November 12, 1982.
[12] "Buchanan Issues Stiff Challenge on Vet College," *The Guardian*, December 14, 1982.
[13] "Lee Readies Case to Get Vet College," *The Guardian,* January 15, 1983.
[14] Richard Hatfield, speaking at the dedication of AVC, May 9, 1987.

Chapter 6

[1] A second well was added after the college opened.

Chapter 7

[1] The relative cost of running veterinary colleges appears to be an irritant at many universities. In October 1986, Dr. R. L. West, American Veterinary Medical Association director of scientific activities, wrote in the *Journal of the American Veterinary Medical Association:* "One hundred faculty members for 400 students is an alarmingly high number to many university administrators who are accustomed to seeing one history professor lecture to 400 students at one time, and deans of colleges of veterinary medicine must explain repeatedly that lecture courses are only a part of the veterinary program because so much instruction takes place in laboratories and clinics."

Chapter 9

[1] AVC teaching hospital.
[2] Atlantic Provinces Economic Council, *The Contribution of the Atlantic Veterinary College to the Economy of Atlantic Canada*, October 2000.
[3] Reg Thomson, *Summary of Considerations Concerning an Atlantic Regional Veterinary College*, January 1979.
[4] Animals were housed at the college and at a farm in Winsloe, north of Charlottetown.
[5] An equine ambulatory service, headed by Dr. Ian Moore, was initiated in August 1996.

[6] G. Edward MacDonald, *If You're Stronghearted, Prince Edward Island in the Twentieth Century*, Prince Edward Island Museum and Heritage Foundation, 2000.

[7] Atlantic Provinces Economic Council, *The Contribution of the Atlantic Veterinary College to the Economy of Atlantic Canada*, October 2000.

[8] Alex B. Campbell, letter to Pierre Elliott Trudeau, September 9, 1977.

[9] R. G. Thomson, *Summary of Considerations Concerning an Atlantic Regional Veterinary College,* January 1979.

[10] Atlantic Provinces Economic Council, *op. cit.*

[11] In later years, he used a similar system to conduct short courses at other universities in the region.

[12] Fish Health Unit *Annual Report,* 1989.

[13] Atlantic Provinces Economic Council, *op. cit.*

[14] *AWC News*, No. 9, Summer 2002.

[15] Atlantic Provinces Economic Council, *op. cit.*

[16] The results of the study had not been made public at the time of writing.

Chapter 10

[1] Spears, Annie, *An Assessment of the Expenditure-based Impact of the Atlantic Veterinary College of the University of Prince Edward Island on the Local Economy*, Department of Economics, UPEI, August 1995.

[2] Spears updated these figures in November 2000. She estimated that AVC then generated between $47 million and $52 million within the provincial economy. Total tax revenues accruing to the province as a result of income generated by AVC's operation amounted to between $8.5 million and $9.5 million.

[3] Atlantic Provinces Economic Council, *op. cit.*

[4] Ibid.

Chapter 11

[1] John P. Brown and Jon D. Silverman, "The Current and Future Market for Veterinarians and Veterinary Medical Services in the United States," *Journal of the American Veterinary Medical Association*, No. 2, July 15, 1999.

[2] Margaret R. Slater and Miriam Slater, "Women in Veterinary Medicine," *Journal of the American Medical Veterinary Association,* Vol. 217, No. 4, August 15, 2000.

[3] George Guernsey, Paul Doig, Peter Fretz, and Diane McKelvey, *Veterinary Medicine in Canada, Opportunity for Renewal*, June 1998.

[4] Guernsey, Doig, Fretz, McKelvey, *op cit.*

[5] Quebec had frozen its tuition fees at 1990 levels, and its students enter the veterinary program at an earlier age.

[6] Slater and Slater, *op. cit.*

[7] Brown, Silverman, *op. cit.*

[8] Guernsey, Doig, Fretz, McKelvey, *op. cit.*

[9] Guernsey, Doig, Fretz, McKelvey, *op. cit.*

[10] Brown, Silverman, *op. cit.*

Bibliography

Bibliography

Books

Clark, Andrew H. *Three Centuries and the Island.* Toronto: University of Toronto Press, 1959.

Dunlop, Robert, and David Williams. *Veterinary Medicine: An Illustrated History.* St. Louis, Mo.: C.V. Mosby 1996.

Ells, Dale. *Shaped Through Service: An Illustrated History of the Nova Scotia Agricultural College.* Truro, NS: Agrarian Development Services Ltd., 1999.

Forbes, E. R., and D. A. Muise. *The Atlantic Provinces in Confederation.* Toronto: University of Toronto Press, 1993.

Gattinger, F. Eugene. *A Century of Challenge: A History of the Ontario Veterinary College.* Toronto: University of Toronto Press, 1962.

Harris, Robin S. *A History of Higher Education in Canada.* Toronto: University of Toronto Press, 1976.

Ings, Jayne, Blaine MacKinnon, and Wendy Rowell. *The History of Veterinary Medicine on Prince Edward Island.* Charlottetown:1983.

MacDonald, G. Edward. *History of St. Dunstan's University 1855–1956.* Charlottetown: Board of Governors of St. Dunstan's University and PEI Museum and Heritage Foundation, 1989.

MacDonald, G. Edward. *If You're Stronghearted, Prince Edward Island in the Twentieth Century.* Charlottetown: PEI Museum and Heritage Foundation, 2000.

MacLean, Angus. *Making It Home: Memoirs of J. Angus MacLean.* Charlottetown: Ragweed Press, 1998.

Paton, Valerie K. *History of Veterinary Medicine in Prince Edward Island.* Charlottetown: University of Prince Edward Island, 1981.

Swope, Robert T. *Opportunities in Veterinary Medicine Careers.* Lincolnwood, Ill.: VGM Career Horizons, 1987.

Waite, P. B. *Lives of Dalhousie University.* Montreal: McGill-Queens University Press, 1994.

Whelan, Eugene, and Rick Archbold. *The Man in the Green Stetson.* Toronto: Irwin, 1986.

Reports

Archibald, J. A., T. J. Holland, and H. M. Vance. "Veterinary Manpower in Canada." Report of the Commission of Enquiry to the Council of the Canadian Veterinary Medical Association, July 1977.

Atlantic Provinces Economic Council. "The Contribution of the Atlantic Veterinary College to the Economy of Atlantic Canada." October 2000.

Atlantic Veterinary College. "Accreditation Information for Council on Education, American Veterinary Medical Association." AVC, University of Prince Edward Island, September 1983.

Atlantic Veterinary College. "Accreditation Self-Study Report, Atlantic Veterinary College, University of Prince Edward Island." November 1988.

Atlantic Veterinary College. Curriculum Survey of AVC classes (various years).

"The Atlantic Veterinary College Interprovincial Funding Agreement, 1996–2001." June 1997.

Council of Maritime Premiers, The. "Intergovernmental Co-operation Among the Governments of New Brunswick, Nova Scotia and Prince Edward Island." July 1984.

Fish Health Unit Annual Reports (1986–1993).

Howell, D. G. "Report of a Study on the Establishment of a School of Veterinary Medicine in the Atlantic Region." Maritime Provinces Higher Education Commission, 1975.

McAllister, Ian (Ed.). "Working with the Region." Henson College, Dalhousie University, 1997.

Spears, Annie. "An Assessment of the Expenditure-based Impact of the Atlantic Veterinary College of the University of Prince Edward Island." Department of Economics, UPEI, August 1995.

Symons, T. H. B. "To Know Ourselves. The Report of the Commission on Canadian Studies." Association of Universities and Colleges of Canada, 1975.

Thomson, R. G. "A Discussion Paper on the Proposed Atlantic Regional Veterinary College." April 1979.

Thomson, R. G. "Summary of Considerations Concerning an Atlantic Veterinary College." January 1979.

Newspapers, Magazines, Newsletters

Atlantic Veterinary College Newsletter, AWC News, The Daily Sentinel Review, The Guardian, The Toronto Star, The Globe and Mail, Island Harvest, Atlantic Insight, The Chronicle Herald and Mail-Star, Farm Focus, The Daily News, University Affairs, The Island Farmer, Atlantic Fish Farming, Journal of the American Veterinary Medical Association, The Canadian Veterinary Journal.

Interviews

Dr. Natasha Cairns, July 18, 2002
Dr. Tom Wright, July 19, 2002
Dr. Suzanne Kennedy, July 18, 2002
Dr. Sylvia Craig, July 19, 2002
Dr. Sherri Coldwell, July 19, 2002
Wayne MacMillan, July 22, 2002
Bennett Campbell, Aug. 13, 2002
Graeme MacDonald, Aug.19, 2002
Cliff Campbell, Aug. 19 and Aug. 26, 2002
Dr. Dale Paley, July 19, 2002
Dr. George Irving, July 18, 2002
Dr. Clayton MacKay, July 19, 2002
Dr. Andrew Robb, July 2, 2002
Michael Murphy, May 22, 2002
Dr. Ray Long, July 19, 2002
Joe Clarke, Nov. 13, 2002
Bob Nutbrown, Nov. 18, 2001
Helen Thomson, Nov. 16, 2001
Jim Lee, Dec. 6, 2001; July 26, 2002
Emery Fanjoy, Dec. 10, 2001
Eugene Whelan, Nov. 19, 2001
Dr. Herb MacRae, March 8, 2002
Dr. Robert Curtis, March 1 and 18, 2002; May 28 and 20, 2002
Dr. Ross Ainslie, March 8, 2002
Shelley Ebbett, Oct. 17, 2002
Dr. Caroline Runyon, Oct. 11, 2002
Marion MacAulay, Oct. 1, 2002
Dr. Alice Crook, Oct. 8, 2002
Dr. Ronald Baker, Feb. 10, 2002
Dr. Peter Meincke, Oct. 15, 2001
Glenn Palmer, Sept. 19, 2002
Dr. Larry Heider, April 15, 2002
Dr. Ken Ozman, Nov. 13, 2001
Dr. Claudia Lister, Nov. 4, 2002
Pauline MacDonald, May 9, 2002
Heather Cole, May 17, 2002
Cora Conrad, May 17, 2002
Mel Gallant, April 10, 2002; July 5, 2002

Dr. Alton Smith, May 16, 2002
Clive Stewart, May 20, 2002
Henry Phillips, Nov. 9, 2001; July 8, 2002
Dr. Lawson Drake, Sept. 5, 2002
Dr. Wendell Grasse, Oct. 11, 2002
Jim O'Sullivan, May 16, 2002
Dr. Jane Hogan, Sept. 4, 2002
Dr. John Anderson, June 11, 2002
Irwin Judson, Aug. 4, 2002
Dr. Ronald Dunphy, June 10, 2002
Reg. Gilbert, May 7, 2002
Larry Durling, May 16, 2002
Barry Stahlbaum, Oct. 28, 2002
Dr. Larry Hammell, Jan. 27, 2003
Dr. Rick Cawthorn, June 5, 2003
Dr. Jeff Davidson, June 5, 2003
Dr. Pierre-Yves Daoust, Oct. 29, 2002
Wade MacLauchlan, Oct. 29, 2002
Dr. Willie Eliot, Sept. 9, 2002
Douglas Boylan, Feb. 28, 2002
Dennis Olexson, Jan. 13, 2003
Dr. Jim Bellamy, Jan. 13, 2003
Jean McDonald, Sept. 4, 2002
Dr. Art Ortenburger, Oct. 17, 2002
Dr. Gerry Johnson, July 8, 2002; Sept. 9, 2002
Dr. David Weale, July 5, 2002
Dr. Andrew Peacock, Nov. 1, 2002
Alex Campbell, Oct. 23, 2001; Feb. 10, 2002
Dr. Tim Ogilvie, July 22, 2002; Oct. 17 and 28, 2002
Dr. Bud Ings, May 10, 2002
Jim Lee, Dec. 6, 2001; July 2, 2002
Andy Wells, Oct. 29, 2001
John Buchanan, March 9, 2002
Barry MacMillan, Nov. 29, 2001

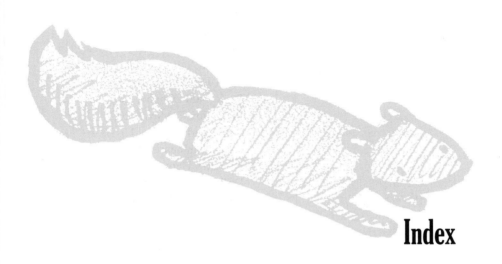

Index

A

Acadia University 27, 32, 33
accreditation 52, 85, 89, 95, 116–7,
    146, 172
acupuncture 183
Adams, Dr. W. M. 55
agriculture 29, 56, 177
Agriculture Canada 22, 31, 80, 97
Agriculture Canada Research Station
    (Wolfville) 33
Ainslie, Dr. Ross 19, 21, 24, **24**, 169,
    189
Akins, Dr. J. T. "Joe" **20**
alternative medicine 183
Alzheimer Disease 123, 125, 126
ambulatory service 132
Amend, Dr. Jamie **93, 94**, 97
American Veterinary Association 85, 89,
    91, 117, 146
anaesthesiology 134, 135
Anatomy and Physiology, department of
    94, 97
Anderson, Dr. John 32, 38
Andrews, Theresa **138**
animal health legislation 20
animal health technicians 134
animal pathology labs 20
Animal Productivity and Health
    Information Network (APHIN) 144
animal shelters 158, 181
Animal Welfare Unit 132, 146, 158–9,
    163, 174
aquaculture 34, 74, 76, 80, 144, 147,
    149, 153, 174, 177
"aquamarine college" 62, 65, 74–76, 80
aqua-sciences 127, 147
Aquatic Animal Facility 154
Aquatic Diagnostic Services 154
aquatic life 83, 149
architecture 83–89
artificial breeding units 20
Association of American Veterinary
    Medical Colleges 146, 172
"Atlantic College of Veterinary Medicine
    and Animal and Marine Centre" 68
Atlantic Fish Health Inc. 154
Atlantic Innovation Fund 152
Atlantic Provinces Chamber of
    Commerce 57
Atlantic Provinces Economic Council
    (APEC) 57, 170

Atlantic Veterinary and Aquamarine
    College 74–76
Atlantic Veterinary College
    advisory council 80
    budgeting 87, 109, 171
    construction 48, 55–56, 57, 83–101,
    115
    cost of 83
    dean (first mention of) 43
    economic spin-offs of 48, 170
    Equipment Replacement Fund 137
    expansion 172, 173
    first mention of 33
    funding 40, 41, 49, 50, 56, 68–71,
    78– 80, 109, 171, 216
    hiring for 97
    mandate 131, 163, 194
    negotiations for 60–72, 77–80
    official opening **70**, **98**, **111**, 120,
    132
    operating agreement 79
    relationship to UPEI 88, 103, 106–
    118
    renovations to 172, 173
    reputation of 131, 174
    salaries 109–111
    sod-turning 88
    vision for 47, 147, 155
AVC Inc. 146, 154–5, 172

B

Baker, Dr. Ron 28, **28**, 29, 32, 34, 41
Barnes, Donna 157
Barr, Michael **132**
Barrie, John **102**
Basher 157, 158, **158**
Beaton, Glen 85
Bellamy, Dr. Jim **93, 94**, 97, 99
Bernardo, Theresa **112**, 116
Binns, Premier Pat 148
Bland, Dr. Michael 23–24
Briand, Heather **96**
Brown, Carol **177**
Buchanan, Premier John 42, 51, 55, 56,
    57, 61, **61**, 62, 66, 67, 68, 69, 70,
    72, 76, 77, **88**
Burka, Dr. John **93**, 97

C

Cairns, Dr. Natasha 179, 180
Campbell, Bennett 74, **76**, 80, 88
Campbell, Cliff 82, **84**, 85, 87, 91, 92
Campbell, Gordon **98**, **100**
Campbell, Premier Alex 26, **27**, 33, 39, 72, 147
Canada, government of
    funding from 57
    role of 40, 65
Canadian Agricultural Services Co-ordinating Committee 22
Canadian Aquaculture Institute 146, 154, 155
Canadian Co-operative Wildlife Health Centre 155
Canadian Faculties of Agriculture and Veterinary Medicine 172
Canadian International Development Agency (CIDA) 40, 78, 99
Canadian Journal of Comparative Medicine, 46
Canadian Veterinary Medical Association (CVMA) 23, 51, 52, 85, 146, 166, 172, 182, 188
Canadian Veterinary Medical Association (CVMA) *Journal* 95
Cardigan Fish Hatchery 146, 154
Carnegie Foundation 29
Carr, Bennett 69
Cawthorn, Dr. Richard 97, 152
Centre for Marine and Aquatic Resources 154
Cepica, Dr. Arnost **93**, 97
Chappell, Prowse 75, **76**, **98**
Charlottetown, city of
    support from 34
Charlottetown Veterinary Clinic 141–2
Chisholm, Don 133, **134**
Cole, Heather **93**
Clarke, Joe 74, 76
Class of 1990 106, 108, 110, 114, 116, 121, 145, 176, 177
class size 17
Climo, Lindee 178
Clough, Dennis **84**
Coldwell, Sherri 180
Cole, Heather 121
Companion Animals, department of **94**, 97, 137
Conference of Atlantic Premiers 50

Connell, Peter 74
Conrad, Cora 89, **93**, 122, 123, 126
continuing education 132
convocation 116, 123, 126
Cornell University 46
Council of Education of the Canadian and American Veterinary Medical Associations 146
Council of Maritime Premiers 32, 40, 49, 50, 54, 64, 69, 72
cows 14,15, 21, 142–3, 149, 176
Craig, Dr. Sylvia 77, **78**, **108**, 115, 176
Cribb, Dr. Alastair 162
Crook, Dr. Alice 159, **159**
curriculum 85, 89, 106, 133–4, 181–2, 183
Curtis, Dr. Robert "Bob" 46, 47, 89, 92, **93**, **94**, 95, 97, 123, **124**, 125, 128, **132**, **134**, **138**, 141, 142, **184**, 185

D

Daley, Evelyn **93**
Daley, Joanne **145**
Dalhousie Medical School, 31, 34, 35, 37, 56
Dalhousie School of Dentistry 35, 56
Dalhousie University 23, 29, 31, 33, 34, 36, 37, 38
Dalziel, Verna Lee **139**
Daoust, Dr. Pierre-Yves 156, **156**, 160, 179
Darris, Joan **93**
Davidson, Dr. Jeff **138**, 153, 155
Davies, Terry **84**
Dean Reginald Thomson Fund 127
Dean's Council 123
Derry, Sharon **93**
Diagnostic Chemicals Ltd. 148
diagnostic service 20, 132, 169, 174
Diagnostic Services 148, 153
diagnostic tools 134, 151
Dingee, Dr. Carl 167
Dohoo, Dr. Ian **93**, 97, **138**, 144
Dohoo, Dr. Susan 93
domoic-acid crisis 152
Doughart, Barbara 48
Downe, Don 88
Drake, Dr. Lawson 106, 111, 112
drugs, veterinary 19, 20, 21, 141
Duke 130–1

E

Ebbett, Shelley **93**, 97, 99
Edinburgh Veterinary College 16
Eliot, Dr. C. W. J. "Willie" **98**, **100**, 103, 106–110, **111**, 112, 113, 116, 117, **124**, 125, **132**
enrolment 55
environmental health 159–61, 185
Epperly, Dr. Elizabeth 146
equine treadmill 135, **136**
equipment 83, 91, 93, 99
European veterinary schools 16, 17, 23
Evening Patriot [Charlottetown] 80
exhibition hall 92

F

Faculté de médicine vétérinaire 17
fallout shelter 91–92
Fanjoy, Emery 50, 64
farming
    crisis 29
    medicine 14
    in Prince Edward Island 29, 124
    service 19, 20, 21, 22, 93, 116, 141
Farm Service Group 138
fast-tracking 87
fees 14, 19, 20
Fisher, Dr. George C. 20, **20**
fisheries industry 56. 152
fish health 54, 62, 127, 174
Fish Health Unit 92, 101, 116, 131, 145, 149, 150–1, 153
Five-Party Agreement 76, 77, 82
food-animal service 22
foreign students 101
fourth veterinary college in British Columbia 54, 55, 57
fourth veterinary college in Canada
    support for 23, 24, 31, 50, 51, 52, 57, 62, 65, 147, 176, 188, 189
Friends of the Christofor Foundation 159

G

G. Murray and Hazel Hagerman Scholarships 167
Galbraith, John Kenneth 29
Gallant, Mel 88, 89, **93**, 120, 122, 127, 137, 146, **184**, **186**
Genome Atlantic 172
Ghiz, Premier Joseph 92, **98**, **100**, 112, 125
Giffin, Ron 50
Gormley, Sharon **100**
Goudie, Joe 75, **76**
government co-operation 72
graduate degrees 34, 109, 112, 117, 127, 146, 169, 171
graduate students 167, 173
Graduate Studies and Research 97
Grant, Dr. Kenneth 64
Grasse, Dr. Wendell 135, **139**, 141, 182
Greater Charlottetown Area Chamber of Commerce 65
Griffin, Rita **114**
Guardian, The [Charlottetown] 33
Guelph, Ontario 14, 16
Guinea Pig Stew and Other Recipes 115
Guy, Dr. Norma **160**, **163**

H

Hagerman, Verna B. 167
Hammell, Dr. Larry **138**, 153
Hancock, Dr. E. Errol I. **20**
Hanic, Dr. Louis 151
Hatfield, Premier Richard 61, **70**, 72, **98**
Hayman, Roland 51
Health Management, department of **94**, 95, 97, 130, 153, 172
Heider, Dr. Lawrence "Larry" 117–8, **118**, 126, 146, **147**, **184**, 187
herd health 141, 143, 174
Herriot, James 187, 189
higher education in Atlantic Canada 49, 50
Hill, Dr. Brian **94**, 97, **104**, **110**, **124**, 125, **132**, 137, **137**
Hitt, Nancy **145**
Hogan, Jane **100**, 101, 115
Hogan, Dr. Patricia **136**
Holland College 167
honorary degrees 124, 126, 128
Horse Infirmary and Veterinary Establishment 16
horses 14, 21, 93, 138, 140
hospitals, human 134, 135, 141
Hovingh, Dr. Ernest **113**, **138**
Howatt, Erwin **132**

Morgan, Gary **98**, **111**
Morgan, Mose 35
Mount Allison University 27
Moynihan, Dr. W. A. 22
Murphy, Mike 99, 101
mussels 144

N

National Research Council Institute for
 Nutrisciences and Health 174
neutering 21
Newfoundland and Labrador Veterinary
 Medical Association 23
Newfoundland Provincial Veterinarian
 20
Nielsen, Ole **184**, 185
North Carolina State University 55
Nova Scotia
 educational institutions in 35
 government of 33
Nova Scotia Agricultural College
 (NSAC) 23, 29, 32, 36, 38, 51, 54,
 66, 68, 71, 79, 80, 169
Nutbrown, Bob 66, 67, 74, 75, **75**, 76

O

Ogilvie, Dr. Tim **93**, **96**, 97, 121, **122**,
 127, 128, 130, 137, 172, **172**, 173,
 178, 180, 183, **184**, 189
Olexson, Dennis 148
Ontario Veterinary Association 52,
 54–55
Ontario Veterinary College 14, 16, 17,
 19, 24, 33, 36, 43, 46, 51, 52, 55,
 95, 101, 153, 169, 172, 177, 189
open house **162**, **186**
ophthalmology 134
Ortenburger, Dr. Art 133, **136**, 183, **185**
orthopaedic surgery 134
outreach 163, 169
Ozmon, Ken 71–72

P

Paley, Dale 101, 114
Palmer, Glenn 26–27, 48
Pathology and Microbiology, department

of **94**, 97
Paynter, Geoffrey **96**
Peacock, Dr. Andrew 167, 182, 188, 189,
 190
Peckford, Premier Brian 61, 72
pharmacists 134
Phillips, Henry 50, 67
"Pioneers' Wall" 126
"planning co-ordinator" of AVC 41–42
politics, role of 39, 60–72, 171
Pork Production Innovation Group 154
preventive medicine or health care 48,
 127, 141, 143, 153, 189
Prince Edward Island
 balance of payments for 49
 benefit to 38, 169–70
 Director of Veterinary Services 20
 farming tradition in 29
 lifestyle 95, 110, 114
 Provincial Veterinarian 97, 127, 172
 Veterinary Medical Association 17
Prince Edward Island, government of
 premier's office 47
 support from 34, 39, 41, 55
Prince Edward Island Equine Retirement
 Society 159
Prince Edward Island Humane Society
 159
Prince of Wales College 26, 27, 29
Pringle, Dr. John **136**
public health 185

Q

Queen Elizabeth Hospital 99
quotas 22, 24, 54, 77, 99, 171

R

R. G. Thomson Academic Achievement
 Medal 78
radiography 135
radiology 92, 134, 137
recession 57
referrals 133
Regan, Premier Gerald 40, 41, 42
registered nurses 134
religion, role of in PEI 26–27
research 132, 174, 206–12
 grants 116, 213